Science and the Educated Man

Science
and the Educated Man

Selected Speeches of Julius A. Stratton

With a Foreword by
Elting E. Morison

The M.I.T. Press
Massachusetts Institute of Technology
Cambridge, Massachusetts, and London, England

Foreword

In the last chapter of this book Julius Stratton tells how he first came as a student to M.I.T. with the idea that he "would stay a little while" and then go out to do some of those things in the great world that he found "so immensely appealing." It did not work out that way. In spite of many opportunities and exterior attractions a "compelling force" held him at M.I.T. for over forty years, and over these years the strange mixture of feelings he had known as an undergraduate developed into "a consuming devotion."

Some part of this book, therefore, has to do with the object of this devotion—the extraordinary life of M.I.T. in the last four decades. To this subject Dr. Stratton brings not only the knowledge acquired as faculty member, Provost, Chancellor, and President but a remarkable array of other capacities. He has the physicist's ability to handle abstract ideas, the historian's sense for the workings of passing time, and something of the philosopher's concerns for what it all adds up to. All these talents are needed, and come through, in the pages in this book that are devoted to the development of M.I.T.

Because this development occurred at a time when the changing intellectual foundations of the Institute were also becoming increasingly the intellectual foundations of our western culture, most of the things Dr. Stratton has to say go far beyond the description of a single institution. Growing up at M.I.T. with its special set of intellectual dedications, he obviously found himself thinking increasingly about the world that was being given so much of its shape by the ideas and energies of science and engineering.

This book has, therefore, many wise observations about the nature and place of science in our culture, many sensible remarks on the topic of C. P. Snow's "two cultures," and many illuminating points to make on the differences between science and engineering. Indeed, one of the most interesting and valuable aspects of the pages that follow is the author's continuing concern to get things straightened out in the matter, now so often confused, of what is science and what is engineering and how each affects the other. Though a physicist by training and at heart, Dr. Stratton is also "a Tech man" continuously attracted by the excitements of putting ideas to work in an engineering way.

He is also, as was suggested earlier, a good many other kinds of man—or at least a man with sensitivities and perceptions in many different fields. So he is constantly running beyond his own professional knowledge into other areas of thought and action. The result is that he has written here about most of the significant elements that go into the making of our modern society—from painting to management and from the place of the individual to the nature of a university in the modern world. And, though the parts of this book were written for many different purposes over a span of several years, the author has consciously or unconsciously produced a remarkable unity in the whole. It comes from his persistent desire to bring all the varied elements he deals with, all the diverse perceptions he has, into some kind of connection; to make some kind of sense out of the modern condition. In so doing he has not produced an all-inclusive, neatly dovetailed, logical system, but he has described most of the essential things we have to think about in trying to order our present experience and he has given us useful and wise ways to think about them. It is a considerable achievement.

ELTING E. MORISON

Contents

vii

CONTENTS

PART II

M.I.T.: INSTITUTIONAL AND PERSONAL EXPERIENCES

PART I

SCIENCE AND THE EDUCATED MAN

1

Science and the Educated Man

A speech delivered at the American Institute of Physics on February 2, 1956.

In the summer of 1900 the city of Paris celebrated the arrival of the twentieth century with a great exposition. That was a year of almost universal hope and confidence in the future, and accordingly the exposition was directed toward a Golden Age of science and industry. Among the 39,000,000 visitors who passed by those exhibits was an American by the name of Henry Adams, and in a famous autobiography he has recorded his thoughts on this preview of things to come. Henry Adams was the product of a tradition of unity and stability. Under the conflicting forces of the nineteenth century he had seen that unity breaking up, and he was searching for what he called a dynamics of history that would anticipate the changing course of mankind. In the Gallery of Machinery at Paris he thought that at last he had found a solution in science.

It is curious to read in the final pages of his autobiography his account of the new Daimler motor, and of the automobile "which, since 1893, had become a nightmare at 100 kilometers an hour." He tells of radium, X rays, and the Branly coherer. For the first time he saw "frozen air" and the electric furnace. The dynamo impressed him most, and it seemed to Adams that among the thousand symbols of ultimate energy the dynamo was not so human as some but it was the most expensive. And then in a prophetic paragraph he wrote of the new American: "the child of incalculable coal power, chemical power, electric power and radiating energy, as well as new forces yet undetermined—will be a sort of God compared with any former creation of nature. At the rate of progress since 1800, every American who lives into the year 2000 will know how to control unlimited power. He will think in complexities unimaginable to an earlier mind. He will deal with problems altogether beyond the range of earlier society."

More than fifty years have passed since that was written, and the progress of our efforts to master the physical universe continues to accelerate. Science is the key to that mastery. No other influence is acting today with comparable force to transform the character—indeed, the very foundations of American life. The products and processes of science dominate our industry and determine our economy, they affect our health and welfare, they have altered our role in the family of nations, and they will govern the conditions of war and peace.

No one can remain impervious to these changes; but it seems to me that the total magnitude of change and the gathering momentum of material progress still are not fully understood. The world into which we were born is gone; we have little or no idea of the world into which our children may grow to maturity. It is this rate of change even more than change itself, this transition from the stability of the Victorian era to some new future state of equilibrium whose shape we cannot as yet foretell, that I see as the dominant fact of our time.

We are asked to look forward, to discern as best we can what the future may hold in store for science, and to anticipate the needs of this new world that science is creating. This is a period of exciting discovery in physics. I imagine that forty years ago Rutherford and his associates worked with the same tense expectation as their experiments began first to disclose the structure of the atom. Now this excitement is shared by hundreds, or even thousands, of physicists working in laboratories all over the world. In a few short years we have had to abandon our simple notions of matter. Elementary particles are appearing in bewildering number and array. We now enjoy resources for the study of physics such that we on earth can duplicate conditions that prevail in the stars themselves. Day by day new discoveries are reported that reveal how tantalizingly close we may be to a comprehension of the very nature of matter itself. If I were competent to do so, it would be tempting to speculate on the implications of these discoveries for the future development of physics and to forecast the direction in which the search may lead.

We might also discuss the probable influence of these advances in physical science upon our technology. The recent progress of science has accumulated an enormous capital inventory upon which technology has as yet to draw. Henry Adams intuitively foresaw the rising abundance of available energy as the primary source of technological change. Our whole complex industrial civilization would be immobilized without mechanical and electrical power; but the usefulness of power is determined by a capacity to control and adapt it to our ends. Through electronics our ability to measure, to compute, to predict, and to control is growing fabulously from year to year, even from day to day.

It might be rewarding to consider how the mounting speed of transportation and ease of communication are shrinking the globe. Literally and figuratively, as John von Neumann has recently pointed out, we are running out of room and at long last have begun to feel critically the effects of the finite size of the earth. There are countless other developments on the technological frontier which it would be interesting, but not fruitful, to examine. For the particular achievements of science and engineering seem to me of little import in the shadow of the great unsolved problem of how we shall learn to live in harmony and prosperity in this new world of our creation.

As we look to the future the questions of how and whom we

shall educate transcend all others. The attitude of our people toward education, the plan and philosophy upon which we conduct it, will determine in large measure whether this new generation will exploit science and technology for good or for evil, and whether knowledge will continue to advance.

Since I have chosen to speak of education before this gathering of physicists, you may think that I ought to consider first the needs of science. There is an appalling shortage of qualified teachers of science in our secondary schools. Instruction in physics, chemistry, and mathematics appears to be losing rather than gaining ground at the secondary level; and yet, in the face of these limitations on the supply of young scientists, the national need continues to grow because of the new stress in industry on research and development, and the apparently insatiable demands of our huge defense program. These are critical problems and they deserve attention. But I believe they are symptomatic of a deeper trouble rather than fundamental in themselves. With effort we can patch up the worst defects of high school instruction; we can perhaps encourage a somewhat larger number of young college graduates to take up the teaching of elementary science and divert a larger proportion of youth into the fields of science and engineering. It is urgent that we make this effort, and quickly; but such measures are at best expedients—a nibbling at the edge of a greater problem which leaves the heart untouched.

For we must view science in the perspective of the broad culture of the country. The education of scientists cannot be isolated from the educational aims and patterns of our population as a whole. The ranks of science and engineering—indeed of all the professions—rise out of the total body of cultivated people. The aims and modes of thought of the young men and women who flow into the professions, the common foundation of knowledge upon which every profession must rest, all this is limited and predetermined by the cultural horizons of the American public. And so I believe that we must look to the roots of *all* education in the American school and college. In this respect our concern will be no less than that of the lawyer or doctor. We cannot hope to alter greatly the regard for science or the quality of its instruction in our high schools until we have dealt with the more fundamental defects in the aims and processes of the high school itself. And in our more mature years we, as scientists, cannot live in isolation from the remainder of the community. We are a part of the total culture, and we may be sure that science in this country will enjoy respect and support only to the degree that its purpose and methods are understood by the public at large. It would be idle to insist that every citizen become sophisticated in the ways of science. But the intellectual temper of a nation is set by that small group of people whom I shall call educated men. There are indeed scientists among them, but I am concerned with that greater number whose careers will lead them into other fields. They have been "liberally" educated in a traditional design, a design that is largely lacking in the substance of science. I fail to see how one can examine "liberal education" as it is still commonly conceived in the United States without concluding that it has lost relevance to the problems of our day.

Because I believe that soundness and balance in the design of liberal education are vital to the future of our science, I shall want to explore the matter more fully. Let me, however, revert for a moment to a particular set of attitudes which affect primarily our high schools but in no small degree also affect our colleges.

We Americans seem instinctively to take delight in the devising of efficient systems to accommodate large numbers. I am in fact sure that the American genius lies in an extraordinary power of organization for production. The whole economy rests on this concept of large numbers with emphasis on moderate quality at low cost rather than on goods of fine substance at a higher price. This genius is the source of our strength but, likewise, a symptom of our weakness. The weakness is compounded when quantity

7

production becomes the single goal or aim of educational planning.

In our democratic concern to solve the problem of equal opportunity for all, we are tending to ignore the need to provide special opportunity for some. In our preoccupation with size we are losing the perspective of quality.

Certainly I do not mean that we have achieved equality of educational opportunity in the United States. The recent White House Conference made it abundantly clear how far we still are from attaining that goal. The normal barriers to progress in education are either economic or intellectual. Our task is to eliminate every economic barrier to advancement, but we should acknowledge more frankly that intellectual limitations are inherent in the human mind. I think that we tend to ignore these limitations, to wish them away. The college degree itself has become the important objective, not so much the advantage of going to college for those who are intellectually qualified. We must not seek to lift the median level of public education by slicing off the peak. By a sharp focusing of attention on the requirements and limitations of the average individual, we are failing in our responsibility to the most gifted. Only by meeting this obligation to the most talented of our children can we hope to maintain leadership among nations. In a contest of numbers alone we shall surely lose in the end. In no field is this more certain than in science and engineering, which are the key to our future security and prosperity. The time has come to speak out boldly and eloquently on behalf of excellence, excellence even at the sacrifice of quantity.

There is, to my mind, a second defect in the processes of the American educational system which is related to the first. In this country we tend to perpetuate in the university the attitudes and character of the secondary school. In large measure I believe this is the consequence of our national inclination to average down the standards of higher education in order to accommodate all those who aspire to a college degree. Too many students are entering our colleges expecting to be taught rather than determined to learn, and they are fortified in that illusion by the attitude of the faculties themselves. The whole familiar process of corrected problems, of weekly quizzes, and midyear grades adds up to an almost intolerable burden, an academic overhead, so to speak, of vast proportions. It is difficult to reconcile that burden upon a faculty with the maintenance of scholarly excellence. But that in itself is not the crucial point. By the prolongation and intensification of secondary-school experience into the undergraduate years of college we weaken or destroy intellectual initiative; we forget that the development of intellectual self-reliance is more vital than the accumulation of factual knowledge; we fail to keep pace with the maturing mind of the student.

I cling to the belief that a university should be more than an extension of high school, that at that level a faculty should teach as much by example as by instruction, and can render no greater service than to convey the meaning of scholarship and to instill a sense of high aims of mind and spirit.

And now let me turn to the substance of liberal education. First I shall remind you that in every section of the country our technical schools have shown an increasing concern for the role of the humanities in the education of the engineer. This is a movement that reflects a growing awareness of the dominant part that scientists and engineers must play in the future of our society. It indicates an appreciation of the full status of a profession and its responsibilities. In almost every engineering school one may observe a liberalization of the curriculum. On the whole I think that we may be content with these developments. It is well to add history, philosophy, and literature to develop powers of judgment and taste and to give balance to the whole. However, we must not be beguiled into believing that a sprinkling of the humanities is the key to culture. The educated man is distinguished by an attitude toward learning and a method of thought rather than by any particular domain of knowledge. Engineering education will attain professional stature by a concern for principles, a ceaseless

9

reaching down for fundamentals, and a rejection of the immediately useful in favor of a painstaking search for understanding. The attitude which we, as teachers, take toward the subjects of science and engineering can have a more significant effect in opening and freeing the mind of the student than can philosophy taught without spirit. Our engineering schools have become conscious of their defects, but they are beginning to perceive also their latent resources in strength. And I am confident of their future.

Whatever its shortcomings, education in science and engineering is on the move. I see no grounds for an equal confidence in the present state of the liberal arts. From the nontechnical colleges will come that body of educated men whose judgment and understanding must largely temper the public attitude toward science. Yet, in a world that increasingly will be dominated by science and its products, it appears to me that liberal education has failed to keep pace with the changing character and expanding needs of the society which it should support.

Let me distinguish sharply between the ideals of liberal education and its current practice. These ideals are indigenous to Western civilization; they do not alter with the times. A liberal education is designed to enlighten, to impart a love of knowledge and wisdom. Its essence, according to Whitehead, is an education for thought and for aesthetic appreciation. It purports to deal with human values, with problems that are timeless. It undertakes to prepare the student to read, to listen, to appreciate all that is lasting of man's works in art, music, literature, and thought.

But a liberal education must also be relevant to time and circumstance. It is an education for cultivated men in every walk of life, and it should fit them to perceive and comprehend the great issues of our time, the forces that are shaping our destiny. It is my belief that modern man must take full account of the role of science and technology. We may draw upon the past for principles to guide our conduct and art to stir our imagination; but the liberally educated man must comprehend the best that is known and thought about the world in which he lives and the laws that govern the material universe.

Sir Richard Livingstone, that very distinguished advocate of humanistic studies, remarks that this past century has witnessed a steady attrition of standards in the sphere of ideas and a gradual breaking up of a philosophy of life which has been accepted in the West for 1,500 years. The two chief instruments in this break-up he considers to be the otherwise beneficent forces of liberalism and science. He conceives of freedom and reason as "the chief forces of liberalism. The liberal believes in freedom for its own sake as giving the fullest opportunities to the human spirit, as encouraging and enabling its self-development, as alone adequate to its natural dignity and powers." And of science, he says its method "is to ascertain facts, to grasp them accurately, and to find explanations for them . . . it is a training in observation, in precision, in objectivity and in a rational habit of mind." But the main limitation of natural science, according to Livingstone —and I quote him because among our contemporaries he has defended most eloquently the traditional cause of the liberal arts —is that science is not human, whereas we have to live with human beings—including ourselves—and nearly all the problems of life are human, while the problems and subject matter of physics, chemistry, and biology are not.

Here we really come to the root of the matter. For now, and increasingly in the time to come, we are destined to live not only with ourselves but with the problems and with the products of physics, chemistry, and biology. It is inconceivable that we shall continue to understand either ourselves or our relations with one another if educated people remain in their present ignorance of the nature of science.

The need, as I see it, is not one of replacing the liberal arts by science but rather of restoring a proper balance. The study of science, or let me better say of our physical universe, ought to be undertaken in our liberal arts colleges with the same thoroughness and serious purpose that once were devoted to Latin and

Greek. The "history and philosophy of science" is not science. A lecture or two on the "scientific method" will impart little understanding of the true nature of the process. "General education" in so far as science is concerned is not the answer.

Perhaps I can best convey my thought by recalling to you the character of that classical education that for centuries molded the minds and set the standards of Western men. In medieval times, apart from the learning of scholars, education for the great mass of people was limited and specialized in guilds and trades, and through this specialization it was highly divisive. And yet there were binding forces of a culture that imparted a remarkable degree of homogeneity to all Western society. This inheritance that we call the classical tradition set the modes of thought, provided the language not only of intellectual leaders but of kings and statesmen down into the nineteenth century. It was a tradition drawing upon the mythologies of antiquity, the philosophies of Greece, and the laws of those very practical Romans. Ultimately the names of real and mythological characters and events were fused into a language of reference and allusion by which educated men communicated with one another. The full sense of Milton, or Keats, or Shelley appears only in the light of classical learning. Generations of British statesmen spoke and thought in these terms. A classical education served as the great vehicle of liberal thought.

We in the United States have lost that facility for classical reference derived from an intimate familiarity with the antiquity of Greece and Rome. A common bond and a common language that served educated people for more than one thousand years has disappeared with the classical tradition. I do not think that we shall soon see it revived.

There is one great, unifying force working in our age, and that is science. We must turn to science for the lingua franca of modern men and find in science the vehicle of modern thought. The legends of antiquity have for centuries provided the symbols and structure of ideas. The gods of Olympus and the heroes of

Troy became human as they were invested with human qualities through great poets. Science is woven through the fabric of modern life and already today is beginning to supply new themes and symbols to art and literature.

Thoroughness and unity of purpose were virtues of a classical education that we might do well to preserve. Perhaps some measure of that thoroughness may be restored through science. The entire burden cannot be left to the college. Education in science must be rooted in the elementary school and developed in our high schools with a seriousness of attitude and method that rarely exists today.

Let me now, finally, pull together these first thoughts on what I feel to be one of the great cultural issues of this coming age. Among all other problems I place first the education of our youth. As scientists we are properly concerned for the profession of science. There are defects in the education of scientists which we must remove. We are alarmed by the growing discrepancy between the need for scientists and engineers, on the one hand, and the visible supply, on the other. We are troubled by the dwindling number of qualified teachers of science in our schools, the lack of incentive and prestige in the whole field of teaching, the seeming failure of science to draw its due from the most gifted of our youth. We are rightly concerned, too, because of unaccountable but manifest signs of open hostility toward science and scientists in many quarters of public opinion.

These are vital matters for each of us, but I think that the remedy lies outside the profession itself. All the outer forms and even the inner forces of our contemporary civilization are molded and controlled by science and technology, and yet we have failed to make the understanding of science a part of our common culture. Anti-intellectualism, as it is called, the thinly veiled hostility to the "egg-head," is the inevitable symptom of a distrust of the unknown. "Electronic brains," nuclear weapons, and guided missiles are modern magic, and their potentialities for evil are as real as though in fact they had been contrived by the devil. We must allow no gulf to grow between scientists and the great body of educated people. The education of scientists and engineers is now too serious a matter to remain wholly the concern of the profession itself. The liberal education of all people is a matter of equal moment to us as scientists. In our generation the classical tradition has lost meaning and relevance. It contained values and standards that we must preserve in the new tradition of scientific learning that is now in the making. The age in which we live may provide man's greatest epic. We have in our hands the power to destroy ourselves or to survive in unity, in peace, and in prosperity.

2

Into New Lands:
The Principles of Navigation

The principal address at the Forty-fourth Commencement of the Rice Institute, May 31, 1957.

Because of its high ideals, the Rice Institute has long enjoyed the respect of the entire academic world. Because, too, of my own warm affection and great esteem for Dr. Houston, your president, I responded with pleasure to his invitation to join you this evening.

Dr. Houston and I met first in the winter of 1927–1928 as postdoctoral students of physics in Munich, Leipzig, and Berlin. To you who are about to graduate from college, a span of thirty years must seem to have its roots buried in antiquity. To us, and to the older members of this audience, 1927 was only yesterday. And something of a case may be made for each of these points of view.

Certainly it is true that in the long perspective of history, thirty years is an almost imperceptible interval. But events have transpired in these *particular* years which have changed the tempo of time and the very meaning of its measure. The great river of human affairs seems suddenly to have come upon a stretch of steep falls and cataracts and is now rushing on toward a new world unlike anything man has known before. Never before our time has the character of our globe as a habitation for man been altered so much in so short a span. From this point of view, 1927 was indeed a long time ago.

The year 1927 stood almost midway between the two most devastating wars of history. I am not competent to analyze the factors that have contributed to these terrible conflicts of our century nor to assess the responsibility. The immediate causes clearly have been economic and political.

But underlying these convulsive movements of our time has been the striving of great peoples to adjust to a new existence made possible by the onrushing development of science and technology. Within this single century man's power to direct the forces of nature both to useful and to destructive purposes has multiplied a thousandfold. Within the lifetime of you who are graduating here this evening there has been an almost explosive growth in the technology at our command. So swift indeed has been the rise of science and engineering as mighty shapers of our society, and of all civilization, that many still fail to comprehend the breadth and strength of their influence. Let me give some substance to this point.

By its triumphs over disease, science has notably increased the normal expectancy of life and thereby affected the balance of populations. It is estimated, for example, that one person out of every twenty who have ever lived is alive today. The effective dimensions of the globe have been shrunk by the ease and rapidity of transportation and communications. Our generation has seen a truly stupendous expansion of industry, both in the volume of business and, even more spectacularly, in the variety of products. These in turn have contributed enormously to the material comforts of life. Science in agriculture has multiplied the fertility of our land and is effecting a major readjustment between town and country. City planners, in fact, are predicting that by 1975 the population of our metropolitan regions may be 70 per cent as large again as it is today, with a resulting total urbanized area three times its present extent. It is difficult at this moment, consequently, to foresee the ultimate form of the

American city; we can be sure only that it will differ vastly from the city of today.

Finally, I would remind you of profound changes that have begun to affect the processes of industry and commerce. Indeed, we are on the verge of a second industrial revolution far more dramatic than the first. Old trades will disappear, of course; but for every job lost a dozen new ones may emerge, in service as well as production, challenging human skill and imparting variety and new interest to human labor.

In these few words I have hoped to convey to you only a suggestion of the magnitude of the revolution that is being wrought through progress in science. I well realize that it has become almost a commonplace to speak of technological innovation in this way. But despite all that has been written and said, I am obsessed by the thought that there is to this day still too little comprehension of the total extent of the changes in which the world is engaged.

There are, of course, in all this history of change, fearful implications for our future. The other evening at dinner I heard a distinguished diplomat remark with deep emotion and sincerity that the world would have been a better place had the atom never been discovered. He thought it now most unlikely that his children would live out their natural lives. But the atom does exist, has always existed, and by manifest destiny man was bound to discover it. The atom of itself is not evil. Neither is it good, although science—in everything it has given us—has increased man's opportunities for good. But the power of good lies with man himself and with his capacity for wisdom and moral judgment, his will to exploit his intellectual and spiritual resources for the welfare of mankind.

I am reminded of the old legend of Gutenberg and the invention of printing. It is told that one day as Gutenberg stood looking down upon his type with pride in his great discovery, the devil appeared before him and in a vision revealed to him all the evil that was to befall man in ages to come through the printed page—the slander and calumny, lies and obscenity, incitement to riot and war. In a fury of remorse Gutenberg caught up a hammer and was about to destroy his precious type. But there came to him at that moment an angel, and in another vision he saw the greatness of books as the record of all that man has said and thought and done that is noble or true or beautiful. Each man might now turn to his own Bible to seek consolation and hope in times of trouble; in history he might read the lessons of the past; in the rhythm and imagery of poetry his spirit would be stirred by the beauty of this world; and in philosophy he might discover the deeper meanings of life.

As it has been with printing, so may it also be with the atom and with the multitude of momentous discoveries that science is now showering upon us. You should view the prospect not in a dark mood of alarm and despair but in optimistic terms of opportunity and challenge. There are Gargantuan problems to be resolved, it is true, before the world can achieve political, economic, and social equilibrium; but I, for one, am confident that they will be resolved and that the difficulties and complexities of your existence will be more than offset by the challenge of new powers. Indeed, I envy you the intellectual adventures that lie ahead. We need waste no time in regretting an age that is now gone forever but must look rather to the future.

Whether this future is to prove better or worse than the past will depend wholly on our own efforts. We can only be sure that it will be different. We have been caught up in a great stream of technological development and are being swept forward into new lands. We cannot stem that tide. I do believe that we can navigate. Indeed, I have faith that we *can* direct the forces of destiny and in significant measure determine the shape of things to come. Without that faith and confidence, your outlook would be most somber.

Now the basic resource that must sustain us in this adventure into the future is the common education of our people. In fact it has been said that civilization is a race between education and catastrophe. This week from schools and colleges all over the land, the Class of 1957 moves out to take its place in the line. The success with which *you* deal with the problems ahead will be governed in large measure by whatever wisdom my generation has exercised in the planning of your education. You in turn must shortly assume responsibility for those who come after. The welfare of this nation—and indeed the peace of the world—will be profoundly influenced by your concern for the importance of higher learning.

At every level of the American school system we are confronted today by perplexing and difficult problems. There is a shortage of buildings to accommodate the surging growth of population. There is an appalling shortage of teachers qualified to meet our more exacting demands. Among educators themselves there are deep differences of opinion as to the function of the common public school, differences as to the appropriate nature of the curriculum or the process of teaching.

At the root of all our concern, moreover, is the fundamental dilemma of numbers—how to reconcile our dream of universal opportunity in education with a pressing need to cultivate excellence. It is imperative that we learn to take better account of the enormous spectrum of human ability, interests, and initiative. Truly, democratic education imposes upon us the duty to help each individual make the most of his abilities. We must recognize, therefore, the obligation of a democratic people in the interest of its own survival to seek out and educate the notably gifted, the industrious, and the ambitious, the most promising of its human resources for spiritual and intellectual leadership.

These are among the most meaningful and difficult questions of our day, and with this Commencement they become your problems too. Since you cannot escape them, I trust you will seek to resolve them with all the intelligence at your command.

And so from these allusions to broad issues let me turn to comment briefly on engineering and the arts in the context of a liberal education for our time.

Engineering education as it commonly exists today is a product of the mid-nineteenth century. It was designed for a more elementary technology and a simpler society. As such it has provided a rather rudimentary training in mathematics and the physical sciences, a meager amount of general education, and the essentials of some specialized branch of engineering. More often than not the emphasis has fallen upon the vocational rather than upon the fundamental or theoretical. While this plan has been basically sound for the years gone by, the complexities of modern technology and the complete transfiguration of modern society confront us with an entirely new range of requirements. The traditional patterns of engineering education are in large measure outmoded and inadequate.

Our first task, it seems to me, is to take better stock of the vast variety of activities encompassed under the title of engineer. When we refer to the alarming shortage of engineers, we have in mind a wide range of occupations that begins with research and development and extends through design and production, into sales and service, and on to the highest levels of industrial management. Every part of this spectrum can have dignity and importance, and each contributes in a particular way to the national good. I do not believe, however, that one and the same plan of engineering education can serve both the aims of immediate utility and the highest aspirations of the profession.

The time has now come, in fact, when the engineering educator must differentiate more clearly and firmly between the engineer and the technologist. Because the great problems of engineering can no longer be dealt with in isolation from the needs and affairs of people and human institutions, I am moved by the vision of a new and higher role for the engineer in the changing world I have portrayed for you. But if the engineers of the future are to provide the leadership that we shall desperately need, then technical competence is not enough; they must also be educated men. At the same time, it has become manifestly impossible within the brief undergraduate span both to train young men and women for technical competence and to educate them for the grave responsibilities of leadership.

Education is a costly investment in time and money, and we must not add lightly to that burden. But for the engineer of the future there seems to be no alternative course. No longer can we afford to substitute mere mastery of techniques for professional competence.

And now, finally, to these comments on engineering education I would add that our need to strike a new balance in the teaching of the liberal arts is, if anything, even more acute. As we have come to require the scientist and the engineer to understand the social and cultural forces they affect and are influenced by, so must we expect the liberal arts student to understand the spirit and meaning of science. Science has become an intimate and influential part of our culture, and no one born into today's world can escape the impact of science on the changing pattern of civilization.

The health of our national economy is bound up tightly with new products of research. Our national security is almost wholly dependent upon them. The books and journals of the day are preoccupied with the political and ethical consequences of an expanding wealth of power through technology. Even art and architecture are affected by the new ideas and new materials that flow from scientific and technical progress. These are the conditions that exist, and we had best learn to live with them.

After all, if we are to contribute constructively to the preservation of peace and the dignity of human existence, then we must be prepared to deal intelligently with the world as it is. If we are to be effective in life, we must comprehend the great issues of *our* time and the forces that are shaping *our* destiny. While we may draw upon religion for codes of conduct and upon history for principles of action, every liberally educated man or woman must know something of the laws that govern our material universe.

This is not to ask that the doctor, the lawyer, or the banker should learn and retain the details of science. But it does seem to me vital that at some period every educated man and woman should have gained an insight of certain principles, certain methods, certain great modes of thinking that lie beneath all science. It is not necessary that a man be able to discuss the gravity of the fall-out problem in terms of roentgens. But it *is* important that he should have a feeling about how we have arrived where we are. He should have explored some science far enough to have gained confidence: confidence that in this great domain there are no mysteries accessible only to the few, confidence that he himself might have pursued science to its farthest frontiers. Out of such confidence and familiarity may develop judgment, an intuitive sense of the critical factors entering into great issues.

Out of confidence, too, and through this basic understanding must come freedom from the fear of science, the dispelling of that mistrust and dormant hostility that threatens to cut off the community of scientists and engineers from the general body of educated people.

This growing chasm in our culture has come about in part because educated men have failed to communicate adequately with one another. Too often the scientist and the engineer have been content to lead parochial lives; too often they have been too little concerned that others should understand the meaning and implications of their work. But this spirit of separatism has arisen too because we live in an age, as one physicist has put it, "in which poets and historians and men of affairs are proud they wouldn't begin to consider thinking about learning anything of science, regarding it as the far end of a tunnel too long for any wise man to put his head into." Such a tunnel philosophy, which is more common than many suspect, should lie heavy on the conscience of American education.

We need desperately a new and healthy fusion of learning and knowledge in our American colleges—a fusion which the educational programs I have outlined in such sketchy terms this evening could help achieve. Happily, I believe such programs are coming. Already we have taken steps to give science and engineering students more opportunities for study in the social sciences and the arts. Happily, some schools are also wisely requiring that the liberal arts student acquire a basic understanding of mathematics and science. But such forward-looking programs are still too few in number to give American education the significance and relevance it needs to create a climate of mutual understanding and respect among educated men. Such a climate, and such a greater unity of purpose in our educational programs, can contribute immeasurably toward the attainment of our highest national aspirations. The challenge of attainment is yours.

3

The Role of the Engineer

An address presented to the Twelfth M.I.T. Alumni Regional Conference held in Washington, D.C., on March 1, 1958.

We have been hearing a great deal about science lately, and it has been our good fortune that science in this country has now found extraordinarily eloquent spokesmen. Thanks to their efforts, there is a growing comprehension of the crucial role of science in our society, of the reasons why it is imperative that the teaching of science in our schools be improved, and why there is a most pressing need for the encouragement of basic research.

Now the true aim of pure science is to know and to understand. Without an understanding of the principles that underlie the raw data of nature, progress is slow and empirical. But to *know* is not enough. If science is to be more than an academic pastime, men must also *do*. To do—to translate into tangible benefits the advances of science—is the function of the engineer. Science and engineering together constitute a vast and continuous spectrum of human effort. No one part of that spectrum, from the ultraviolet of basic research to the infrared of manufacture, can flourish or even subsist without the remainder.

It is my impression that in our concern to stimulate science we may be overlooking, and indeed neglecting, the role of the engineer. And so it is on his behalf that I should like to make these few remarks. For engineers are not merely the draftsmen and plumbers of science. Our airplanes, our missiles, our great weapons systems, the stupendous growth and development of industrial processes are all the direct products of American engineering.

If now the public confuses engineering with science, if the press last month hailed Wernher von Braun as a top "missile scientist" rather than the excellent engineer that he is, the blame must be borne in some measure by the engineering profession itself. It seems to me that engineers have been notably inarticulate in proclaiming the high mission of their profession. One cannot escape the conclusion that, as a group, engineers to some extent are failing to grasp and to respond to the challenge, the opportunities, and the responsibilities that are rightfully theirs. On the one front where engineering meets science, physicists, chemists, and mathematicians have shown themselves seemingly better able to cope with the vastly complex and difficult problems of component and systems development. And on that other front that joins men with machines, one may observe that social scientists are rapidly moving in.

Why is this so?

First, consider for a moment the technical origins of our industrial power. This industrial might of the United States has come about because we have achieved the highest rate of productivity per worker and because we have become a nation of people accustomed to technology.

It is the American engineer who has been the chief architect of this industrial development. But the technology with which he has dealt has, until very recently, rested far more heavily upon the practical arts than upon science per se. Steelmaking, glassmaking, textile manufacture, and the whole gamut of our basic industries arose and grew with a minimum of formal technical training on the part of their leaders.

By contrast, new industries are today springing up on every hand—indeed, almost overnight—as a result of some new specific scientific discovery. To my mind, the most significant economic factor of the mid-twentieth century lies in this proliferation of totally new products and processes, coupled directly to the forward march of science. Here are the real and healthy roots of a developing economy and a second industrial revolution.

For one hundred years, the education and training of the American engineer has been meshed and in tune with the expanding industrial complex.

Suddenly, largely under the impetus of World War II, the whole scene has changed. With great skill and perception our industrial and military leaders are learning to capitalize upon discoveries in this wondrous new world of science. The routine work of engineering technology must and always will remain; but day by day—almost hour by hour—are added new tasks which challenge not only the ingenuity but also the intellectual and educational resources of a totally new kind of engineer. The demands upon the engineering profession are soaring ahead; the horizons are lifting and widening.

We need engineers in quantity, as you well know, to maintain the rhythm of the national industry. But where are the men to come from—men with the depth of professional knowledge and the width of view—to deal with these great new tasks?

They will come, of course, as they have in the past from the schools and colleges of engineering throughout the country. And the question that is pressing upon all of us who are entrusted with the engineering education of America is whether we, as educators, have opened our eyes and fully comprehended the great destiny of this profession.

Let it be said to the credit of the engineering institutions that they are clearly awake to the need of a reappraisal. There is scarcely an engineering faculty in the country that is not currently restudying its curriculum, reviewing its aims, revising its methods. Basically the issues are one and the same for every school; if now I seem to speak more particularly for M.I.T., that is because my own responsibilities lie there, and there I see the problems most clearly. To this let me add that the very issues now so hotly debated in the educational world have their counterparts in some of the most perplexing problems of industry and government.

First, may I remind you that engineering is itself many things, a vast domain of many lands. It includes the highly skilled art of the modern technician; it encompasses managerial occupations requiring a consummate knowledge of human behavior; and it has been pushed forward now into areas of systems design—with its array of problems that can be mastered only by the most sophisticated methods of modern science and mathematics.

Every part of this extended domain contains elements essential to the industrial life of the nation. Our great professional societies, such as the institutes of mechanical and electrical engineers, choose to group these activities together in a few large categories. But the fact is that within any one of these traditional branches there is a vast range of professional qualifications. The term "electrical engineer" no longer describes adequately the occupation. Nor can one and the same plan of engineering education be used for all.

One of the major criticisms of our lower school system has been the unwillingness to distinguish among students of various aptitudes, ambitions, and intellectual gifts. The professional schools must face the same necessity. I believe that our engineering institutions must begin to redefine their objectives more clearly and design their curricula accordingly. We, too, have our problem of "roles and missions."

Second, the dissolution of the boundaries that traditionally marked off one domain of engineering from another gives rise to an array of new and perplexing questions. Within, let us say, electrical engineering, the range of activities grows wider and wider. At the same time the electrical field is fusing into the mechanical, and it becomes increasingly difficult to distinguish one profession from the other.

Engineering principles and engineering systems cut boldly

27

across these arbitrary barriers, and professional education—particularly at the undergraduate level—must follow suit.

It becomes increasingly clear that no matter what departmental structure we may cling to, we must also find better ways of breaking freely out of this compartmentalization. Almost every institution is experimenting with various devices, such as interdepartmental laboratories, research centers, and seminars, as a means of providing new syntheses and pulling together in a new common effort whole groups of disciplines.

And now, third, we come to the heart of the debate. How much basic science shall we infuse into the education of an undergraduate engineer? How far shall we go in discarding all the drafting and design, the shop and more practical, immediately useful, professional subjects?

Here it is that each institution must make its own hard decision. There are thousands of boys in this country with an aptitude for doing, with a love for mechanical things, and with no mind for mathematics. There is, and there will be, work for them to do—mountains of work—work that is indispensable for the smooth operation of our great industrial machine. It would be folly to cram these heads with advanced calculus and quantum mechanics. Let us not waste and discourage this human talent but rather direct a part of our educational effort—indeed perhaps the largest part—to their particular needs and ours.

But now, in addition, let us also recognize that henceforth the vanguard of engineering, the great creative genius of the profession, will march forward hand in hand with science. The research engineer of the future may be concerned with the anatomy of components or the physiology of systems. His interest may center on materials, fluid flow, combustion, or information theory. Whatever it may be, it is unlikely that in these areas he will make important contributions without benefit of a superb training in physics, chemistry, and mathematics.

As one observes the changing scene at M.I.T., it is apparent that increasingly our faculty is looking to these aspects of the new engineering as our own particular mission. We would, perhaps, be moving in this direction even faster were it not for certain reservations, certain misgivings which I also share.

Because, for all the common ground, engineering is not and never will be science. There is inherent in the profession of engineering a whole set of attitudes and concepts completely foreign to pure science. The engineer must have a feel for materials, a concern for cost, an understanding of the factors of size and weight, an appreciation of the problems of maintenance and replacement, and, above all, an unfailing sense of responsibility toward his client and the public good.

As we move steadily toward the scientific and theoretical in our engineering training, it seems to me of the utmost importance that we preserve the deep qualities of the profession. They are, in a very real sense, the humanities of engineering. I find no conflict in the concept of undergraduate professional training that is both liberal and thorough. That is the kind of an undergraduate experience that M.I.T. aspires to give.

4

The Purpose and Goals of M.I.T.

President Stratton's Inaugural Address delivered at M.I.T. on June 15, 1959.

There is, in my own academic world, no higher honor to which I might aspire. To this honor I add the privilege of following in the path of a friend and colleague of long standing for whom I hold the utmost affection and respect. There is no one who has had a better opportunity than I, or more occasion, to observe at firsthand the tireless energy and devotion with which Dr. Killian has worked for M.I.T., or the enormous contribution that he has made to our country, and I respond to his words with the assurance that for me the prospect of this continuing and intimate association in a common cause fills me with enthusiasm for the task and confidence in its success.

I, myself, come to you this morning as no stranger. First as a student, then through the ranks of the faculty, the productive years of my life have been interwoven with the hopes and progress of M.I.T. These years have given to me a sense of the past —a deep respect for those who have gone before, an appreciation of the thought, the energy, and the devotion that have brought to this institution of ours the high esteem it now enjoys.

But with this regard for the past, I have come to cherish an even brighter vision of the future. M.I.T. is a product of our age. By its aims, its methods, and its ideals it is keyed to the needs and problems of the contemporary world. Today, more than ever, the measure of our greatness will be determined by our capacity to educate for leadership.

The challenge of contributing to such high purpose inspires one to rise above all personal limitations, to approach the task not only with humility and understanding but also with courage and a venturous spirit.

Just fifty years ago, almost to the day, across the Charles, M.I.T. inaugurated its sixth president. Those who listened to Richard Maclaurin on that seventh day of June, 1909—and some are here today—must have been conscious of the stupendous changes that had begun to envelop their world. The Victorian Age was past, and gone with it was a certain stability in many of the affairs of men. Here in the United States especially there was evidence on every hand of expanding material progress, of a rapid rise in the wealth and prosperity of the country. We had become a world power, and Americans were pervaded by a spirit of confidence and optimism.

At the same time, it was becoming increasingly clear that science and technology were the powerful, accelerating forces of our advancement. It is extraordinary, as one looks back, to see what a multitude of inventions that have come to symbolize this modern age were introduced into common use in the first decade of the twentieth century. Just the year before the inauguration of Maclaurin, the United States government signed its first airplane contract. For the sum of $25,000, the Wright brothers agreed to deliver an airplane able "to attain a speed of 40 miles per hour, to sustain flight for one hour, and with the ability to land undamaged." In that same year, Henry Ford introduced his Model T car, and wireless telegraphy had developed to the point that the Marconi Company could open its transatlantic service to the general public.

While any catalogue of the inventions and industrial developments of the early 1900's is impressive, the discoveries of science in that remarkable first decade were, if anything, more prophetic of the future than the advances of technology. Physics, in particular, broke free from its classical mold with the first formulation of the quantum theory of radiation, of the special theory of relativity, and of our ideas on the radioactive decay of the elements. Thus the opening of our century marked one of the great intellectual revolutions of history.

Yet in 1909 this new world was very new indeed, and I can hardly believe that anyone in Maclaurin's audience could have foreseen the fantastic progress of the next fifty years. Today one travels from Boston to Los Angeles in five and a half hours instead of five and a half days. Networks of communication carry our voices, and soon our images, to every corner of the globe. Nuclear reactors and digital computers have become tools of industry and commerce as well as of research. And space, once the lonely outpost of science fiction, is now a new frontier. In short, great engineering and technical developments have advanced our capabilities by many orders of magnitude.

Man has made comparable gains in every field of pure science. Day by day we penetrate deeper into the ultimate mysteries of the nucleus and of the universe. We have created elements and synthesized complex molecules, including biochemical ones. We have developed great experimental tools like particle accelerators and radio telescopes. We can work at the edge of absolute zero and, through thermonuclear fusion, have begun to reproduce the conditions that prevail within the suns. In the span of our lifetime, many of the dread diseases which had afflicted mankind throughout his history have yielded to modern medical science.

But these are only the peaks of great discovery and invention, and their brilliance should not blind us to the massiveness of the developments upon which they rest. For the advances of science and engineering have come to affect every aspect of our lives. They are changing the patterns of our culture and the form of our cities. They are permeating finance and commerce and shaping issues of domestic and foreign policy.

Only as we pause to take account do we discern how far the process of change has carried us and with what gathering momentum we are being swept forward. Yet in spite of all this apparent progress, we can hardly view the future with unalloyed optimism. We have enjoyed an enormous enhancement of material power and wealth in the United States but have notably failed to resolve some of our most urgent social problems within, and we are challenged from without both economically and politically for our very survival. Our hope for a resolution of these problems will depend upon our wisdom and our command of the forces that we have set in motion.

The basic question is, can we in fact control our destiny? I myself have faith that with intelligence and resolution we can. Although I recognize full well that our course as Americans cannot be pursued in isolation from that of the other peoples of the world, nonetheless I am convinced that at this juncture in history the success of our efforts will rest largely in two courses of action, both rooted in education.

First, we must understand that our future economic health, quite as much as our military security, will be governed wholly by our capacity to maintain technological superiority. We have no alternative and must bend all our energies to maintain the advance of science and to expand its frontiers. An inner thirst for knowledge and understanding draws men to research; but it is incumbent upon the universities, upon industry, and upon government to provide a soil and a climate in which research may flourish.

Second, in our concern for external security we must not ignore a wide range of urgent and difficult social problems brought about by the technological revolution itself—problems such as the growth of our cities and population, the interrelation of men and machines, the production and distribution of food, the increase of leisure, to name but a few. In their form, if not wholly in their substance, these are new problems, and the men and women who will deal with them most effectively must have a new kind of education.

It is in this context of national necessity that M.I.T. must examine its role. Very nearly one hundred years have gone by since the founding of the Institute. On this Alumni Day one may look back with a great deal of admiration on the part that M.I.T. has played over the century in developing the industrial power of our country. From our alumni have come men who have helped to construct the highways, the bridges, the great cities. From the earth they have taken the oil and the minerals. They have built and managed great industries. They have been among the foremost leaders of a vast and growing research and development effort in the nation. To a multitude of professions they have brought a mastery of the methods of science and engineering.

As we recall these accomplishments, it is important to remember that this institution was created by William Barton Rogers as an expression of faith in certain new concepts of professional education and that from the very outset our academic policies have been directed by a few central ideas. In essence, Rogers maintained that there is dignity and importance in the mastery of useful knowledge; that the foundations of a professional life may profitably be laid in the undergraduate years, combining with and contributing to a liberal education, to the enrichment of both; and that science and engineering can be the legitimate foundations of a higher education.

M.I.T. has been built upon these convictions. The contributions of our graduates over the years both at home and abroad provide ample proof of their essential worth. I think it well on this occasion that I reaffirm my own confidence in the basic soundness of these principles.

Yet the course upon which they must now guide us leads into a future that will be totally unlike the world of Rogers or even of Maclaurin. M.I.T. must adapt itself to the needs of a changing epoch. It must assume new roles and accept new responsibilities. But as we lift our eyes to ever higher horizons, it must be with the clear understanding that no task is presently more urgent than the education of youth. The greatest contributions that M.I.T. can possibly make to the common good will be made through those young men and women who will have shared with us for a period the experience of striving and learning. Everything that we do, whether for the advancement of knowledge or in the interest of public service, should be viewed in the larger context of our teaching mission. The highest goal to which a university may aspire is that its sons and daughters shall be leaders in art and science and that their influence shall be brought powerfully to bear for the welfare of mankind.

With this affirmation of purpose, I come now to certain thoughts upon the quality of education at M.I.T. and the directions in which we should guide our efforts. There are, in my view, three areas that particularly merit our attention:

First, I think that we must strive to develop more effectively the creative, imaginative, constructive powers of the student.

Second, we must bring about a more productive integration of the humanities and social sciences with the physical sciences and engineering.

And third, I am convinced that we must endeavor to impart to our students a better understanding of the professional estate and of the values that it implies.

Let me elaborate upon these three aims in a little more detail.

Throughout the entire history of the Institute, much of the strength of our educational plan has been derived from the rigor and thoroughness of our method. From the day he enters as a freshman, the undergraduate learns to work in depth and to be held accountable for the results. He learns also to work under pressure and to marshal and employ his knowledge under test. From this discipline and mastery of fundamentals comes an intellectual self-reliance that will stand him in good stead.

We wish in no way to lessen this rigor. But the acquisition of accumulated facts and the formal instruction of lectures and classroom are properly only part of the educational process. The intellectual discipline of tests and problems must be supplemented and enlivened by other forces that will arouse and stimulate the impulses of originality latent in every student.

Some of you may have been fortunate enough to have heard Edwin H. Land two years ago in Kresge Auditorium speak eloquently on this subject. He expressed the conviction that "the freshmen entering our American universities have a potential for greatness which we have not learned to develop fully by the kind of education we have brought to this generation from the generation of the past."

It seems to me that it is in the context of these ideas that research takes on its full and proper meaning in the university. By its very nature, research demands originality in thought and action; and it is in research that the student, as well as the faculty, can find an outlet for creative interest and energy and share in the intellectual excitement of new discoveries. Consequently, university research serves but half its purpose if it becomes remote and isolated from the students themselves.

Of course, I understand that only at the graduate level does a student normally begin to participate effectively in research. I am also well aware of the practical difficulties of undergraduate involvement in advanced work. But I do believe that the spirit of originality and independence of thought that permeates our superb laboratories should begin to influence our students from the time of their arrival. Whether an undergraduate himself produces a piece of work of any novelty is of little moment. What is important is that we stir his imagination, encourage him to break free from the channels of conventional thought, and teach him how to bring to bear upon his problems the facts and methods acquired in the classroom.

As I express these ideas, it is with the conviction that they apply with particular force to engineering education. Lately, engineering has been pushing its roots deeper and deeper into all areas of science and mathematics. This has been a necessary and, indeed, inevitable trend. But we must remember that engineering is art as well as science.

From his earliest history, man has been driven to build and to do, and the fufillment of this urge finds its highest expression in the work of the engineer. The engineer is concerned with making and with producing, with converting the yields of pure science to useful products and services. His function is to adapt knowledge to beneficial ends, to find ways and means of solving the practical problems of human existence. There is, therefore, in the education of the engineer, the most compelling reason to develop by all possible means his creative and constructive powers. The achievement of this goal is one of the great challenges and opportunities in education today.

I come now to my second objective. The contributions that the humanities and social sciences can make to the education of the scientist and engineer have been clearly established. Over the past decade, under the leadership of Dr. Killian and Dean Burchard, the Institute has won wide recognition for the support that has been given to these more liberal aspects of our curriculum. I think it important to say that I, too, am convinced of the wisdom of this course. I also believe that we must now strive to integrate the teaching and research in these areas even more closely with the larger interests of the Institute.

The range of our professional activities at M.I.T. has for some time been steadily widening. We are concerned not alone with science and engineering for their own sake but, increasingly, with fields on which science and engineering have a direct impact in contemporary society. In addition to the obviously related field of economics, we are becoming increasingly active in such areas as psychology and political science. Our Center for International Studies and, indeed, the School of Management also fall into this category. This growth is both desirable and inevitable. However, I feel that our efforts in these new fields will be most fruitful if we are able to capitalize to a greater extent upon our special resources as an institute of technology. In fact, the justification for our excursions into these new areas is that they express a natural extension of the central purpose of the Institute. Although

I am satisfied that notable steps have been taken toward meeting this criterion, much more can be done to bring about a freer and more mutually profitable interchange between the students and faculty of the several schools.

And now third and briefly, we should be reminded that M.I.T. is a professional school, and as such we have an obligation to impart to our students an understanding of both the privileges and responsibilities inherent in the professional estate.

What, in fact, constitutes a profession? In the sense that I am speaking, all the professions share certain qualities in common that set them apart from the other occupations of men. Each, of course, is centered upon a particular field of learning. Each makes high demands upon the intellect and requires a mastery of special techniques. But it is an attitude that distinguishes the professions rather than their particular content. Above and beyond all technical competence, the truly professional man must be imbued with a sense of responsibility to employer and client, a high code of personal ethics, and a feeling of obligation to contribute to the public good.

As a great educational institution we shall fall short of our mission if we fail to inspire in our students a concern for things of the spirit as well as of the mind. By precept and example, we must convey to them a respect for moral values, a sense of the duties of citizenship, a feeling for taste and style, and the capacity to recognize and enjoy the first-rate.

I have ventured this morning to emphasize once again how the extraordinary advances of science and engineering have brought to our contemporary world both new problems and new opportunities. Whatever their solution may be, we shall in dealing with them have to draw heavily upon our resources in education. Because of its character, its traditions, and its achievements, M.I.T. has a major role to play.

Somewhere in his writings, Charles W. Eliot, who was later to become president of Harvard, remarked that when truly American universities appeared, they would be indigenous to our soil, relevant to our time, and would grow out of national need. I can think of no better way of summing up the essential character and spirit of M.I.T.

As I come to the end of my remarks, there are a few final thoughts that I should like to share with you upon the nature and responsibilities of the office that I have just assumed.

A university is an extraordinarily complex organism. It works in many fields of scholarship. It encompasses a wide range of operations involving teaching, research, and, in these days, government contracts. It has obligations to a varied constituency—students, faculty, alumni, and trustees. A university must be administered. As in any great enterprise, there must be a source of prompt, clear-cut decisions and an orderly handling of administrative affairs.

But good administration, indispensable as it is, is only the beginning. It has been said countless times that the faculty is the university. Upon the president himself rests the responsibility of creating and maintaining a climate in which both learning and teaching may flourish. This means an intellectual environment in which imaginations are stirred, which fosters confidence that worth-while things can be done, and where feelings of freedom and security go hand in hand with a sense of obligation and loyalty. In such a favorable climate, president and faculty work together in harmony and share the excitement of planning and building.

But there remains to the president one more function of leadership. In the perpetual debate of ideas that is the essence of a university, he must be more than a referee. He must himself be prepared to take positions on matters of educational import. Above all, he must be able to formulate his aims and make clear what he proposes to achieve. And in all these things he must be guided constantly by a vision of the highest goals of his institution.

To this charge I pledge my whole endeavor.

5

Abstract and Concrete

An address delivered at the Fifty-fifth Annual Meeting of the American Association of Museums in Boston on May 25, 1960.

I feel most highly honored that you should have asked me to speak here on this occasion—and, quite frankly, awed by the prospect. That I have ventured forth at all onto this platform is evidence, in truth, of the irresistible but charming persuasiveness of Perry Rathbone. I hope also that you will consider it evidence of my deep personal appreciation and respect for his contributions, not only as a skillful and imaginative director of our Fine Arts Museum, but also as an outstanding citizen of this community.

A few weeks ago I had the pleasure of a fleeting visit to the lovely town of Williamsburg, and as I wandered about the old streets and onto the campus of the college, I was reminded again of the Virginia roots of the Massachusetts Institute of Technology. For M.I.T. was founded just ninety-nine years ago by a man named William Rogers, who had been a professor at the College of William and Mary, as had his father before him. First at Williamsburg, then at the University of Virginia, later in Baltimore and abroad, Rogers was stirred by the great current of fresh ideas flowing from the nineteenth-century revolution in science and industry. He was enthralled by a dream of education that would unite art with science. Finally, this intellectual ferment of the eighteen fifties drew him to Boston, and here he found a community receptive to his plans.

By a curious chance, it was the changing topography of the city that brought M.I.T. its first association with our neighboring museums. Many of you may know that in those days the Back Bay district of Boston was only a vast expanse of tidewater flats. Most of the area from about where this Hotel Statler Hilton now stands to the present site of the Museum of Fine Arts and northward to the Charles was flooded twice daily by the tide from the river. Then, in 1858, a contract was let to fill the entire Back Bay, and with it came the opportunity to reserve new lands for public purposes—or, as Governor Banks put it, for "such educational improvements as will keep the name of the Commonwealth forever green in the memory of her children"—a statement singularly reminiscent of some of our more recent efforts at urban redevelopment.

To profit from this opportunity, William Rogers was requested to prepare a Memorial to the Legislature—on behalf of a committee of citizens devoted to the interests of natural history, horticulture, the fine arts, and education—to petition for a grant of land in the Back Bay for a Conservatory of Art and Science. That first petition failed, but it foreshadowed the location soon thereafter in the vicinity of Copley Square of the Boston Society of Natural History, the Museum of Fine Arts, the Massachusetts Institute of Technology, and, indirectly, the Boston Public Library and Trinity Church.

M.I.T.'s immediate neighbor on Copley Square was the Museum of Natural History, the antecedent of our present Museum of Science. And, parenthetically, I must confess to my friend Bradford Washburn that the founder of M.I.T. appears to have enjoyed a self-confidence with respect to addressing museum groups that I, his successor, can only envy. For Rogers relates in a letter in 1860 that the Natural History Society proposed to restore "the old usage of a public address" at its thirtieth anniversary meeting that year, and that he was being earnestly pressed to undertake that duty. He tells us that he consented on the condition that his talk should "not be a formal essay, elaborated for publication, but such as I can give without the trouble of writing it." Those, indeed, were the days of self-confident orators!

I have ventured to dwell for a moment on these small bits of local history to recall some early associations that linked a number of our well-known Boston institutions. They serve also to set the theme for my remarks this morning; for I should like to express to you a few thoughts upon the interdependence of science, of art, and of education. Specifically, I want to comment upon the growing movement toward an abstract intellectualism to be observed in all these fields. One may approach this subject from several directions; but I shall feel on safer ground if you will allow me at the outset to use my own institution as a case history.

Most of you are probably unaware that M.I.T. was founded as a corporation—an institute in an older meaning of the word—to embrace three distinct endeavors: a Society of Arts, a Museum of Arts, and a School of Industrial Science.

The idea of a Society of Arts reflected the intense desire of educated men and women in the nineteenth century to keep abreast of the advancing tide of learning. In an age that lacked the dubious diversion of movies and TV, popular lectures were frequently developed into masterpieces of simple and exciting exposition by many of the most brilliant intellectual figures of the day. There still survives at M.I.T. a vestige of the original Society of Arts in a series of Sunday afternoon lectures offered each winter to the public.

The Museum also survives in departmental collections and in the small but distinguished Hayden Gallery. It is significant to observe the major importance that Rogers and the original incorporators attached to this part of their plan. In an early address to the Society of Arts, Rogers reported on his visits to museums at Kensington, Edinburgh, Paris, and Karlsruhe. It was the Polytechnic Institute at Karlsruhe that inspired him to a larger vision of what might be accomplished here in New England. He was struck particularly by the extensive museum of models that were the objects of constant study by students of the school—models of structures, of roofs and arches, of mines and tools, of everything, in fact, that concerned the architect and the engineer.

From these travels and observations came the concepts and the patterns of education at M.I.T. in the first years of its development. The Institute was founded on the idea of "learning by doing"—an idea, as you well know, that was by no means novel to John Dewey. Much of the "doing" in those early days took place in the laboratory, the shop, and the drafting room. Thus, in many respects the original plan of M.I.T. expressed a revolt from the increasingly sterile forms of the classical curriculum that preceded the sweeping revisions in American universities of the 1870's. Pragmatic Americans of that generation responded eagerly to this new emphasis upon the experimental, the visual, and upon tangible ties with the realities of our physical universe. Many of the great innovations initiated in 1869 by Charles Eliot, following his return to Harvard from M.I.T., were undertaken in this identical spirit.

What, now, has happened to this spirit, this philosophy of learning over the intervening century? Despite all the current emphasis upon graduate research, one gains the clear impression in looking back over the years that there has been a steady decrease in the time exacted from an undergraduate in the experiments of the laboratory and a more than compensating increase in hours devoted to the theoretical discussions of the classroom. In making this observation, I am speaking no longer of M.I.T. alone but am pointing out a trend that is basic to education in science and engineering everywhere in the United States.

It is easy to trace, through recent decades, the inroads on time formerly allotted to drafting and other practical exercises, and easy to understand the reasons for them. These encroachments have been offset by a notable rise in the standards set for mathematics and, quite properly, by a growing emphasis upon basic theory in the undergraduate years. This shifting emphasis from the empirical toward the theoretical and abstract has been an essential consequence of a growing maturity in the ideals of American technical education, and it reflects the enormous complexity of contemporary science and engineering. The simple, intuitive methods of attack upon which the engineer, for example, was once accustomed to rely are now wholly inadequate. Only a generation ago a student of engineering had little need for mathematical baggage beyond trigonometry and the most elementary facts of calculus. Today there is scarcely any branch of mathematics, however abstruse, that does not bear directly upon some important group of engineering problems. The process begins at the top and works down. It is first at the graduate level that we find use of the new mathematical tools and the latest abstractions of physical problems. But quickly the new learning and the analytical approach filter down into the undergraduate years, and today they are even exerting a force on the secondary schools. These sophisticated modern techniques, as you would expect—and this is the fact I wish to stress—appeal strongly to our brightest students and, consequently, strongly influence their intellectual temper and attitude.

It will seem strange to some that I, a theoretical physicist by training and by devotion, should voice alarm at this modern trend to the analytical and the abstract. It may seem even more surprising in light of the fact that over the years I have worked increasingly to lift the intellectual and professional standards of engineering education. My concern is that in the process of assuring for the student a mastery of mathematical analysis, we fail to develop equally his other powers of perception. As an undergraduate I spent hours in the drafting room, in shop and foundry, and in various laboratories. I also enjoyed a great variety of industrial experiences, a privilege that will become increasingly difficult for students in the future. In one sense much of this time was time wasted, because I have never since done any drafting, have never poured any more molten metal in a hole in the sand, nor, to my regret, spent time with a lathe. But, in retrospect, it is clear that those hours gave a balance and a perspective to my studies and to my entire outlook. The particular practical courses of my day were perhaps ill adapted to modern needs. They contributed certainly to the stigma of vocationalism in the education of the engineer, and quite properly they have been largely discarded. Nevertheless, they filled a purpose, and it is essential that in some way we find a better means of accomplishing this same end.

Only through action and experiment does a student learn to observe. Analysis divorced from physical objectivity ultimately becomes barren. Whether a boy learns to draft or to paint or to use a tool with adequate professional skill is quite beside the point. What is important is that he gain a sense of the concrete, that he have a direct and tangible experience with the objects and the materials about which he thinks. To the powers of the intellect there must be added the capacity to see, a sense of form and shape and design, a feeling for the plasticity of matter.

It is a curious fact in the record of American scholarship that American mathematicians tend to be the most scornful of the potential usefulness of their labors. It is strange, in the light of our national heritage, that the scholarly emphasis on physics in this country is increasingly on the theoretical at the expense of the experimental. It is significant to observe a comparable movement toward the abstract in the current revisions of engineering education.

There is for each of us a spiritual domain and an ultimate good in pure thought. But man has survived because of a capacity also for ideas that find their final expression in physical terms—in action, in the experiments of the scientist, in the constructions of the engineer, in the creations of the artist and the architect. The power of logical analysis to abstract immensely complicated physical situations constitutes one of the supreme achievements of the human mind, but when abstraction becomes an end in itself, science and art invite sterility.

These are thoughts, I am well aware, upon which reasonable people may arrive at varying conclusions. I have drawn upon my own field of competence to develop the argument but, clearly, I have in mind much larger implications. An almost arrogant intellectualism seems to me to affect a wide domain of American scholarship. Examine, if you will, the teaching of literature in our best colleges. From an emphasis upon an understanding and enjoyment of reading, one observes currently a preoccupation with criticism, with philology, and with symbolism. The stress placed upon the Ph.D. in English, and the increasingly narrow and technical character of the thesis in that field, is evidence, I believe, of this same movement. Philosophy at this moment shows a greater interest in the processes of mathematical logic than in the problems of human existence. I think you will find a corresponding emphasis upon the theoretical and intellectual approach to music in many of our universities. In this audience I had best leave unspoken any comment upon the application of my thesis to contemporary art.

There appears to be a curious dichotomy in our American culture. On the one hand, there is our familiar obsession with material things, with gadgets and devices. On the other, as if in protest or rebellion, our scholarship reveals increasingly this trend toward the analytical and abstract. There is nothing in such an evolution that is of itself insidious or reprehensible. On the contrary, we are entitled to be proud of the new level of American intellectual achievement and maturity. But let us beware of immoderation in this approach to scholarship—of excesses that drain learning of its human content and convey to the student of art and literature and science merely an anemic image of the human drama.

My plea is for balance, for a fullness and roundness in the educational experience. Let the student learn to respect and to cultivate his ties to the physical world, to nature, and to experiment. Let him find in art the counterpoise as well as the companion to the intellect, so that he may learn to see and feel as well as to think. For in the completeness of life human feelings must be added to human thought. Without that completeness, without healthy bonds to art, to nature, and to man, scholarship —however impressive—ultimately becomes little more than an intellectual exercise.

It is in this context that I see the role of the museum. Our museums in all their variety conserve and display our heritage from the past. Thereby they give testimony of the inventive and creative powers of man through the ages, and so transmit these transcendent values from one generation to the next. This is the humanistic lesson of the museum. Because your museums are living instruments of education, you and we of the universities have common objectives—a common responsibility toward the community—and must make common cause together.

6

Personal Responsibility and an Informed Leadership

The Commencement Address delivered at Carleton College, Northfield, Minnesota, on June 3, 1960.

I address these thoughts to the seniors, to you who graduate on this June day. This is *your* day. Commencement is a major milestone in the life of a man or a woman. Formally or informally, in graduate school or in the continuing experiences of a lifetime, the educational process itself carries on. But, with the baccalaureate that you are about to receive, you begin to cast off some of the restraints and some of the privileges of youth. Your energies will be guided by new aims and objectives; and your achievements will be rewarded by a new order of satisfaction. You step out this morning into a world of harsher realities. You must assume your share of responsibility for events that will shape that world in the years ahead.

And what kind of a world will it be? Of one fact only can you be utterly sure: it will be a world unlike any that man has known before; and the problems that even now confront human society are of an urgency and complexity that will challenge all your wisdom and knowledge.

Before I turn to my main theme, let me suggest the most critical of these issues:

First, we in the West must reckon with the almost fabulous

rise of the Soviet Union and the rapid advance of the Chinese Republic. World power is now concentrated in two great rival systems, so that increasingly the peoples of the globe, whether they will it or not, find themselves allied to one or the other of two armed camps.

Second, we must take account of deep political and social currents that are stirring everywhere in Asia and Africa. Old states are awakening, and new states emerging whose experience in self-government is as yet unequal to their high aspirations.

Third, we are faced with dramatic increases in the growth rate of populations that threaten in many parts of the world to outstrip their means of subsistence.

And fourth, I remind you of the awesome and baffling array of political, technical, and biological problems that have followed in the wake of nuclear weapons.

The list has no ending. But these few examples are sufficient to demonstrate that many of the critical problems of our time, as well as the enlargement of our opportunities, are products of the contemporary revolution in science and technology.

A multitude of technological advances has contributed to the growing power of the Soviets. Modern technology nourishes the economic hopes of underdeveloped countries. The population increase is a direct consequence of progress in medical science and sanitary engineering. Our military problem, needless to say, is directly related to the most impressive scientific and engineering achievement of the century—the controlled fission and fusion of atomic nuclei.

Clearly, the dominant forces of our age will result from the impact of new science upon the economic, the political, and the social institutions of mankind. Without some firm knowledge of the substance, the methods, and the *limitations* of science, it will be difficult even for the most thoughtful of men to comprehend with any depth of penetration the forces, the spirit, or the events that are shaping our contemporary civilization.

But science, for the overwhelming majority of men and women, should remain not an end in itself but a means: a means for the amelioration of human existence upon this planet; a means to knowledge and understanding, to mastery of our physical environment; a means, in short, to great good if we are but wise.

As one who by training and experience has been both engineer and physicist, I take enormous satisfaction in the stirring advances of my profession. However, as a citizen I am profoundly troubled by the prospect of a failure to resolve in time issues generated by this very progress. To Americans, that failure can mean the disappearance of our cherished institutions and the loss of our most precious freedoms.

It is difficult, as one surveys the future, to discern in the turbulent movements of the present the specific problem, the particular crisis, that contains the most fertile seeds of our undoing. Yet increasingly of late it has seemed to me that there is one question of overriding significance. The question is simply this: Can a democracy in the American tradition meet and survive the challenge of a highly competitive, highly organized central authority?

The essence of our kind of democracy is a belief in the value of the individual—a belief that each and every one of us has not only the right but also the responsibility to participate actively in forming our laws, in selecting our leaders, in shaping the character of our institutions. Democracy fails when a preoccupation with private privilege leads to neglect of public duty.

The early history of our country affords the model for all time of a working democracy. The emphasis upon personal responsibility accorded with the ardent Protestant thought of the day. The origins of Carleton College should remind you that the Congregational Church had a substantial influence upon the establishment of town-meeting government in Colonial New England. The nobility of these ideas was given eloquent form in our Declaration of Independence and in the Constitution of the

United States. Throughout the history of the republic they have been our guide to policy and action, and for one hundred and fifty years or more they have illuminated the hopes of peoples in many lands seeking dignity and freedom.

What, now, of the present? How successful have we been in amending and adapting our basic democratic forms to cope with the needs of a country that has expanded from the 4 million people of Revolutionary days to the 175 million of 1960?

The stupendous growth of our population, of our cities and states, is by no means our only—or even principal—problem of size. There have been corresponding enlargements in the basic economic units of agriculture, of industry, and of labor. These, in turn, have been matched by an appalling expansion in the units of government. Moreover, the increasing speed of transportation and communications tends to weave the fabric of modern society ever tighter. This grouping of people into larger and more complex units is a normal response to a growing population and to the extraordinary progress of technology.

The impact upon self-government, however, is to subject the processes of democracy to a complete change of scale. In the massiveness of the effort, the influence of individual leadership is diffused and destroyed.

For the past eighteen years I have been a frequent commuter to Washington and a moderately close observer of affairs in the Pentagon. I have learned how easy it is to attribute all error, confusion, and uncertainty to either an alleged stupidity or an unworthy ambition on the part of someone else. There is a common belief among our countrymen that the Pentagon is dominated wholly by a bureaucratic indifference and by interservice rivalries. To a degree these do exist. But, contrary to the public image, it has been my experience that there may also be found in that vast structure countless men of good will— intelligent, loyal, working a length of day that few of us would tolerate in private life, completely dedicated to the welfare of their country. Yet to a shocking extent their best efforts are frustrated; for the situations that normally confront them are of such colossal magnitude that it becomes virtually impossible to understand them in sufficient detail for wise decision, and the mass of the system is so huge that decision more often than not leads to no perceptible action.

And so I believe that among our larger national problems there is no task of greater urgency than that of adapting and developing the machinery of democratic government to the scale and complexity of modern life. Let us put the question boldly. Is it possible, in the spirit of our finest democratic tradition, for men charged with the responsibilities of government to assemble and organize the most essential facts, to agree upon wise decisions, and to achieve prompt and effective action?

If we as citizens are inattentive to the problem of managing our own affairs, or fail to discover appropriate solutions, there can follow, it seems to me, only one or another dismal outcome. One alternative is an increasing mediocrity in the conduct of local as well as national affairs. We shall, in that event, expose growing weakness to the economic and military challenge of highly competitive states under strong centralized authority. Or, in the second alternative, we may find ourselves moving irretrievably toward the same high concentration of federal power.

Either of these courses spells the end of democracy as we have known it, and with that demise may go many of the institutions upon which we have built our American civilization—the freedom of private enterprise, the balance of public and private education, the respect for all religious faiths, the basic liberties of the Bill of Rights.

The word *democracy* has different meanings in different lands. The Russians and the Chinese speak of their democratic processes; but the end of all action focuses there upon the statistical welfare of the massive state rather than upon the respective rights of individual citizens. The hope of preserving our concept of individual freedom lies with us; for if the United States, with the influence of vast economic and military power, fails in its task, we may hardly hope that truly democratic forms of government will long prevail among our Western allies. By our own example, we must show to the world how these great ideas can be made to work.

And now, finally, what is there to do? Remember always that a true democracy implies *duty* as well as the *right* to participate in the affairs of government. Democracy gathers its force from the grass roots. The blight upon public life in the United States today is the willingness of citizens to leave matters to someone else. We grumble, we criticize, we complain; but we make few sacrifices and take little part. We view with alarm the intrusion of the federal government into the affairs of labor and management, into plans for old age relief and medical assistance, into the subsidy of education, and into the support of research—to suggest but a few. But we should ask ourselves honestly whether "leaving it to Washington" has not often seemed an easier solution than first coming to grips with these problems in the local communities.

Civic virtue, in short, begins at home. It starts humbly and in small affairs. It involves everyone. It begins with you who today go out from college. If you want better schools, better government in your town, less poverty and squalor in the city, then give something of your own time and thought to solving the problems of a working democracy. The spread of juvenile delinquency, for example, is a frightful commentary upon the character of contemporary urban life. We shall cure it, if at all, by going to the local roots of the cause—not by federal law or subsidy.

Clearly the larger goal of all our efforts must be to learn how to conduct a truly democratic government on a heroic scale. Either we must make the machine work or we shall go under. There must be innovations in the forms and the processes of government; we must adapt to the uses of government the methods of modern management and engineering. But in the end there will always remain the need for informed, experienced leadership with a willingness to serve. I can think of no higher mission for the colleges of our nation than to develop that leadership. To the Greek ideal of excellence we must add the Roman respect for civic virtue. Carleton has given to you these qualities in trust. I am confident that you will now employ them in the service of your country.

7

Physics and Engineering in a Free Society: A Viewpoint on Education

A speech given at a meeting of the American Institute of Physics, Arden House, New York, on September 29, 1960.

One of the most distinguished novelists of our generation, Sir Charles P. Snow, who by training was a physicist, has been a recent visitor to the United States, and we had rather hoped he might be with us today. His paper on "The Two Cultures" that appeared first in the *New Statesman* and his Rede Lecture, delivered at Cambridge University a year ago, have made an immense impression in his country and ours, because he describes so well a problem that has been a matter of concern to many of us. His subject, you will remember, is the deepening gulf between the literary intellectuals and the scientific community. It is a gulf of mutual misunderstanding accentuated by a breakdown in the means of communicating ideas, by the development of special vocabularies, and, worse, by the use of identical words with utterly different meanings. The rift between science and the arts is acute, although possibly less so in America than in Britain.

However, it would be more accurate to say that in our country today there is in fact only one culture, for science and the arts are completely interwoven in the pattern of our daily lives. This interweaving of the several cultural strands into the single fabric of a modern, western society is quite another matter than the proposal—so often suggested—that there is a basic unity of science and the humanities. This is an appealing concept, but I think it is largely without substance. Science is built upon measurement and a quantitative approach to nature; thereby it is, in its very essence, fundamentally distinct from the great ideas that compose the humanistic tradition. Rather, these two, the arts and the sciences, are complementary parts of a whole. Together, and in balance, they illuminate the modern world.

But unhappily, as Sir Charles goes on to point out in his Rede Lecture, there are signs of cleavages appearing in our society other than that which separates the scientist from the humanist. The stupendous accumulation of knowledge and techniques that marks the twentieth century has led inevitably to a growing specialization in the professions. The consequence is a fragmentation of learning and the intellectual isolation of the scholar, each on his own little professional island. Of course, this sort of thing is by no means wholly new. I imagine that the medieval guildsmen felt much the same way about each other and about the clerical philosophers of their day as Rutherford may have felt about T. S. Eliot in ours. But the consequences of misunderstanding may be more serious to us.

One of these consequences is the failure of the American public to understand the respective roles of the scientist and the engineer and the nature of their contributions to productive industry. And directly related to this matter of public comprehension is the crucial problem of technically trained manpower. What measures must we take as a nation to insure that an adequate number of young men and women enter upon careers in science and engineering? Are the present and foreseeable economic and military goals of the country so endangered by potential shortages that we should adopt a plan of partial mobilization or a system of special inducements as in time of war?

Surely the answer is no. Ours is a technological society, but our highest aim is to preserve it as a free society. Authoritarian states such as China and the Soviet Union appear, at least for the short term, to enjoy an enormous advantage in the planning of their economic development. They can regiment their human resources almost as freely as they can make allocations of minerals or consumer goods. For us, the question of adequate technological manpower is part of a larger issue of whether democracy can survive in competition with a dictatorship. That is a matter upon which we really have no choice. We are unalterably dedicated to the democratic idea; consequently, we must have faith in the power of the democratic process. We believe in the concept of free enterprise, and therefore we must likewise support a free market in the professions.

In this free market, all who represent science and industry—such as we here today—have not only a right but also a responsibility to guide and encourage students in school and college in the choice of careers. We must demonstrate to the very best of our ability the intellectual challenge and excitement of engineering and science, as well as their urgent importance to the national interest. The decisions that these young people in high school or college make about their future professions no doubt will be affected in some degree by the prospect of material rewards—of starting salaries and stock participation. But do not ever underestimate the idealism of youth even in this materialistic age. I am convinced that judgments on the importance, the interest, the intrinsic worth of prospective careers are still the crucial factors that determine each year how many from each fresh graduating class enter the several professions.

The answer to the manpower problem is a deep-seated and intelligent interest in engineering and science aroused early in primary school and continued through high school into college. And how do we stir this interest? By excellence in teaching, and by constant revision and improvement in the content, the techniques, and the methods of presentation.

On this score I take anything but a pessimistic view. Never in American history has there been a more concentrated and enlightened effort to strengthen and advance education in science than at present. The contributions of the Physical Science Study Committee will prove to be a landmark in secondary-school teaching, and I can hardly pay too high a tribute to my colleague Professor Zacharias for his vision and imagination, and for his success in joining in a common enterprise the energies and range of experience of high-school teachers and some of the most eminent physicists from our universities. And the greatest contribution of the Physical Science Study Committee, it seems to me, has been the stimulation it has given to others to do likewise. The time was ripe, and the way was shown. All over the country there is interest and movement. Independent groups in fields other than physics have formed spontaneously and are planning and working. This is a movement that has sprung out of the profession of teaching itself, and we may hope that it will grow in power and range. I am confident that we are on the threshold of revolutionary changes in the teaching of science.

Happily, these developments will not be confined to the high-school level. There are studies in progress that relate to the primary school. Even more significant is the college problem. Not so very long ago professors in our major universities were disposed only to disparage the state of physics in the lower schools. Now they have recognized that we in the colleges must look to our own fences. Here, too, there is a widespread need for revision and modernization. There should be increased emphasis on the aims and evolution of science and on its relevance to our culture. In most institutions, elementary laboratory instruction in physics and chemistry has been notoriously sterile. New experiments must be devised to bring modern physics into the realm of the undergraduate, to excite his interest, and to strengthen his understanding. And finally, there is an opportunity to extend the use of films, modern lecture demonstrations, experimental kits—all the new techniques of teaching that are beginning to show such bright promise at more elementary levels. It is highly satisfying to observe that the American Institute of Physics recognizes its responsibilities in furthering these efforts, and no doubt many of you have read in the September issue of the *American Journal of Physics* the report of a Conference on the Improvement of College Physics Courses which presented a series of recommendations and plans for immediate action.

With these comments on the place of science in contemporary society and with these examples of how education is moving to meet its broader responsibilities, I come now to the principal subject of my remarks—the relation of physics to engineering, and some implications for the education of the engineer.

Just what has happened to the interest in engineering on the part of undergraduates in technical schools and universities all over the country? Everywhere one finds a remarkable shift in enrollment from engineering departments to physics and mathematics.

Let me cite some figures from M.I.T. by way of example. Throughout the five-year period 1956–1960 we have maintained our freshman enrollment essentially constant according to plan. Freshmen indicate their preference upon admission, although they are not asked to commit themselves to any particular professional field until much later. Within the stated period, the preferences expressed for engineering dropped by 30 per cent. In all fields of science there was a corresponding increase, and in physics, in particular, the indicated growth of interest was 60 per cent.

Local conditions, of course, always influence such figures, but I think no one will contest the fact that they reflect a condition affecting schools of engineering everywhere, a condition that is critically serious in its implications for our industrial development.

What is the cause? To what extent has the undergraduate been captivated by the glamour of recent developments attributed, often improperly, to "science"? Many engineers have protested rather bitterly, and with some justification, the practice of our news reporters of crediting to science a number of great accomplishments that are, in fact, triumphs of engineering. Nonetheless, I think it quite clear that the profession as a whole has been rather ineffective in presenting to the public a lucid and compelling view of the true nature of engineering, of the challenge and excitement it can hold out to the brightest minds, of its important and indispensable contributions to society. If we could dig down to the roots of the current shifts in interest among our students, I think we would discover that they stem often from sincere and thoughtful convictions that modern science has become intellectually more stimulating and socially more important than engineering.

For my own part I think this view, should it continue to prevail, could be disastrous for the country. I also think it completely unfounded in fact. But also, in keeping with my earlier remarks, I believe that we must seek our remedy by building upon the inherent strength and appeal of the great engineering professions rather than by resorting to a containment of the current interest in physics through artificial limitations upon enrollment.

We must begin with engineering education itself. We must ask ourselves, more candidly than has been our practice in the past, whether in its traditional forms it does indeed offer a challenge to the very brightest young minds. And does it indeed meet the needs of the contemporary industrial revolution? I think the answer will be no, or at least rarely. But here too, as in the teaching of science, there are signs of great developments in progress.

Let me comment, therefore, upon the conditions that seem to me to underlie the current crisis in engineering education.

Although one looks normally to the academic institutions to provide the innovating forces of education, it is now industry itself, on many fronts, that is already showing the way. But neither in industry nor within the schools of engineering has there been full agreement on the most urgent needs of the profession, nor, consequently, any unanimity of opinion as to how and to what extent our curricula should be revised.

The grounds for this division of judgment are easy to discover. The great bulk of American industry still draws upon well-established, relatively stable technologies. Steel, oil, motors, construction, to name but a few, are the backbone of our national economy. And in these industries the effectiveness of each method or process has been tested by experience. Because of the inherent nature of the operations involved, research has influenced rather slowly the character and volume of production.

To this basic industrial core we must now add the extraordinary and expanding array of new companies whose very lifeblood is research and development. Their field of operations lies along the furthest frontiers of scientific discovery. Their business is to exploit the advances of science, to translate them rapidly and economically into useful products and services. Relatively few of these new enterprises yet bulk large on the industrial horizon. But their role in the current technological revolution, their significance for the future economic strength of this country in the face of rising international competition, are wholly out of proportion to their individual size.

As we go about the task of preparing our engineering students to meet the responsibilities of *their own generation,* we must consider in fair perspective the whole range of future needs and opportunities. Since the middle of the nineteenth century, American engineers, educated in what is now a classical pattern, have taken a leading part in the prodigious growth and achievements of our industries. Any alteration in this proved plan of education must clearly be designed to increase the prospects of further contributions on the part of our engineering graduates.

The argument for revision rests largely on the fact that today's scientific revolution is no mere extension of the industrial revolution. No experienced observer of the contemporary scene can easily escape the conclusion that research and development *are in fact* powerful agents for growth and change. The impact of this innovating force of research bears increasingly upon every segment of our industrial activity—upon our basic industries as well as upon our more novel enterprises. Popular interest in the spectacular success of electronic devices or the prospects of "exotic power packages" should not obscure the greater significance of far-reaching advances on many other fronts. Recent applications of solid-state physics to materials—both metals and nonmetals—and progress in the chemistry and thermodynamics of combustion, in the theory of communications and control, in biochemistry, in the mechanics of fluid flow, to suggest only a few, will in due course affect the operations, the competitive position, and the profits of industry of every size and category.

Industry in the decades ahead will exploit increasingly the progress of basic science. The time lag between scientific discovery and practical application will diminish and the boundary between pure and applied will often be confused. But the achievements of industry, the reduction of ideas and principles to useful products, will remain the work of the engineer. And, as I remarked earlier, one of the important responsibilities falling upon our professional societies, on industry, and our academic institutions is that of conveying to students a clear understanding of the character of the engineering profession, of the challenge and excitement of its opportunities.

The pace of technological change is accelerating. We cannot possibly foresee the progress of discoveries of tomorrow. We ought, therefore, to concentrate our efforts on imparting to prospective engineers a thoroughly fundamental technical competence, together with the versatility and the intellectual self-reliance to keep pace with the advancing frontiers. This means a greater emphasis on basic science; and every conference on engineering education of late has concurred in the need for a higher level of achievement in mathematics, physics, and chemistry.

Until recently our engineering schools had lagged in this regard. This accounts in part, I think, for the rise in physicists in industry. We find them increasingly employed in areas of application and development. One reason surely is that it is they, rather than the engineers, who have comprehended the possibilities for change and innovation. Now I think all of us are quite clear that research and development represent in fact a continuous spectrum of activity. And none of us would overcompartmentalize the work of the physicist and the engineer. It matters not a whit whether a man doing a particular piece of work calls himself a physicist or an electrical engineer. But comprehension of the fact that physics and engineering are indeed different, with different professional missions, is essential.

The pure scientist and the true engineer differ fundamentally in their aims, their motivation, their methods of attack, and the kind of reasoning which leads them to their decisions. In our new emphasis upon fundamentals in engineering education, we must avoid the danger of downgrading other essential attributes of the engineer. There is a need for something more than advanced mathematics, more than an understanding of the systems concept, or the use of operations analysis or optimization techniques. The student ought to have more than a passing acquaintance with a laboratory. He must acquire a deeply ingrained feeling for experiment, for scale and orders of magnitude, and what it means to measure. He needs to appreciate the professional engineer's concern with cost, with reliability, with public responsibility.

Earlier I spoke of the great changes taking place in the teaching of science. The engineering field, by and large, has been slower to take stock of its needs. But I can report to you that there is new life and movement in our engineering schools all across the country, and we can now look forward to a vigorous and imaginative attack on the basic problems.

Let me conclude then with the final observation that these comments on engineering education obviously leave much unsaid. Engineering encompasses a broad spectrum of activities, and there is no single route to useful service in the profession. Whatever the bias of a particular academic institution or industry, we must recognize that there are many valid points of view— views that are complementary, not competitive. And one of the strengths of our free and pluralistic society is that we can and do support a great variety of educational institutions with differing roles and missions. We need technically trained manpower in quantity and quality at all levels. But we urgently need also a creative breed of engineers who will combine a deep knowledge of science and mathematics with the special attributes of the engineering profession itself.

8

The Fabric of a Single Culture

The Centennial Convocation Address delivered at M.I.T. on April 9, 1961.

We have come to the closing hours of our first century and the culminating moment of this celebration. The Massachusetts Institute of Technology is young as measured by the age of many institutions whose delegates have generously assembled on this campus from all over the world to greet us. We are moved by your tributes and by the spirit they convey. I am deeply honored that as President it is my privilege to respond on behalf of the Corporation and the faculty, the students and the alumni, and of the friends of M.I.T. who are gathered here today. I do so most warmly.

This, of all times, is an occasion to look ahead, and it is about matters that lie directly before us that I shall speak to you this afternoon. Yet, for one brief moment, let us cast back our thoughts over the hundred years.

Outwardly and materially, the world of 1861 appears at a backward glance but a crude prototype of our own. Nothing is more obvious than the changes that have been wrought by the stupendous discoveries and countless inventions of the intervening century. Yet the same great currents of thought that we commonly identify with our own time had already begun to stir. The same conflicts of ideas were clearly visible.

Among educated people everywhere, and in the city of Boston

in particular, there was an intense interest in science and invention. The idea of progress had, if anything, a greater hold upon the nineteenth century than upon our own. There was a confidence that science was the ultimate key to the welfare of mankind and a profound belief that in only a matter of time science would reveal to us all the mysteries of our universe, would satisfy the material wants of peoples everywhere, and would bring peace and harmony among men.

Out of these convictions there emerged a vision of the future that stirred the imagination and moved men to attach a new importance to useful knowledge. But there were then as now many who refused to accept this pragmatic view of life. Cardinal Newman in Great Britain, in a celebrated essay, defined the classic idea of a university, maintaining that its true purpose was instruction rather than research—to train the mind rather than to diffuse useful knowledge. Matthew Arnold eloquently defended the humane letters as the single path to culture. Against this ancient fortress of classical learning, Herbert Spencer and Thomas Huxley pressed their case for science and for relevancy in education to the problems of the day. A full century ago the lines of battle between the "two cultures" were already clearly drawn, and the echoes of these great controversies were heard here in Boston.

Jacob Bigelow, a President of the American Academy of Arts and Sciences and a Rumford Professor at Harvard, concluded a memorable address in the 1860's with these words:

> A few years ago, men witnessing the effect of an electric current on a magnetic needle wondered if a motive force could not be transmitted with electric speed to a far distance. A few years ago, men looking at their faces in a glass wondered if such an image could not be fixed on a plane surface, by the agency of light. A few years ago, men toiling slowly and wearily on highway roads wondered if the fatigue and loss of time could not be saved by some better mode of conveyance. A few years ago, men about to undergo surgical operations wished in vain that the attendant pain might in some way be averted. The solution of all these problems

is now achieved by the triumphs of utilitarian science. The nineteenth century, one-third of which is yet to come, has already converted all these wants and wonders into physical and historical facts. Would the recovery of the lost books of Livy, the orations of Hortensius, or the poems of Sappho, be any compensation for the loss of any one of these from among our own cotemporaneous revelations?

The response of the "other culture" to such declarations was immediate and forthright. The pages of *The Atlantic Monthly* in the 1850's and 1860's and the minutes of the several literary clubs of this city record spirited and sometimes acrid discussions of the relative merits of classical and utilitarian studies. They record also the successful efforts of William Barton Rogers to found here a new kind of institution.

M.I.T. is the product of that generation and of a plan founded on faith in the dignity and worth of useful knowledge. Indeed, the history of this Institute over a hundred years is interwoven with the economic and industrial development of the United States. Its progress stems from the need to apply the power and skill of engineering to the growth of a nation.

We of this generation of students, faculty, and alumni of M.I.T. can take rightful pride in the accomplishments of our first century. Yet as we look now to the future, we find ourselves caught up in a great forward surge of science that has no parallel in the past. Hardly a month passes without the announcement of a significant advance on one frontier or another. One would be foolhardy to predict the discoveries and inventions that lie ahead. But there is on every hand a pregnant sense of extraordinary things to come, a belief that man is on the threshold of many revelations, of profound new insights into the nature of life and the construction of the universe.

In our own century science has given to man an unprecedented power over his physical environment. But how shall he govern this headlong advance? How shall he be guided on his course? What shall he ask of science? What shall he do with its products?

How shall *we* as educators teach the generations of scientists and engineers who are to follow?

From these most difficult and urgent questions I draw three thoughts that I should like to express to you today.

First, I want to speak of the advancement of science for its own sake. Science is the great quest for knowledge and understanding of the laws of Nature. Science is a structure that man is erecting piece by piece in the likeness of God's world. The great edifice is rapidly taking form. The beauty of the whole and of its parts is becoming ever more visible. The work commands the interest and consumes the energies of an increasing number of men and women.

Yet, while the quest for knowledge reveals the beauty and harmony of Nature, it can lead also into doubt and into places of darkness and fearful domains.

Just one hundred years ago, the theories of Darwin and Huxley relating to the origin and evolution of species were looked upon by many as an assault upon the foundations of religion. Today, biologist, geneticist, and chemist are pursuing a search that is taking them ever closer to the elements of life itself and may well disclose a complete evolutionary chain that leads from a primordial, inorganic mass into an organic, life-sustaining planet.

The physiologist and the psychologist are making rapid progress with their investigations of the structure and functions of the human mind and personality. As they proceed, there looms before us the terrifying possibility of the evil to which such knowledge may be turned—the power to control and subvert both mind and spirit.

To many people today the idea of a thinking machine, of an electronic device that can simulate at enormous speed the logical processes of thought, is not merely awesome but abhorrent.

And I need hardly remind you of the appalling consequences that may flow from a misuse of our knowledge of nuclear fission and fusion.

One may multiply the examples at will. The prospects of potential evil or disaster are so many and so frightening that thoughtful people question how far we dare proceed. They ask whether science has not begun to trespass upon forbidden territory, whether there are not paths of investigations that henceforth must be barred to man's curiosity.

To this I reply that there is no retreat. We have no alternative but to follow truth wherever it may lead us. One cannot escape evil by ignorance. Whatever preconceived notions we may have of how the world is constructed must and will give way little by little to a more profound understanding of the laws of Nature as they actually are. Knowledge itself, as has been said many times, is neither intrinsically good nor evil; but the power that knowledge gives can be turned to evil purpose. Only our will for right against wrong stands between us and disaster. We must seek salvation not by withdrawal from the quest but in man's own conscience, in his innate sense of decency and morality.

I have been speaking of science as a path to understanding. I want now for a moment to talk of science as an instrument for human welfare.

The entire development of man, from his most primitive origins into a modern, civilized being, is the story of an unfolding interplay between tools and a directing mind and will. There is a common idea that in the long, painful, tortuous evolution of human society the enlarging brain and the progress of intelligence are always in advance of the inventions of technology. On this point the record is by no means clear. In fact, recent archaeological discoveries in Africa appear to link stone tools with prehuman primates more than half a million years ago. Such evidence suggests how behavior through the use of tools may have interacted with changes in anatomical structure, as well as in intellectual development, and how out of these interactions emerged man as we know him.

All our recorded history offers further evidence of this interaction between ideas and technology, between thinking and doing. The great cultural transformation that began to over-spread Western Europe in the fourteenth century was the product of many factors; but any interpretation will be distorted and in-complete that fails to take account of advances in the use of wind and water power or improvements in such simple devices as pumps for exhausting water from flooded mines. The theories of Galileo, Newton, and Huygens are among the grandest works of the human intellect. But the indispensable prelude to these theories was the invention of physical instruments—of telescopes, microscopes, and the air pump. In our own day, the profound new insights into the nature of matter and the constitution of the stellar universe, with all their philosophical implications, are direct consequents of great new machines, of particle accel-erators, and radio telescopes; and these, in turn, have been made possible wholly by progress in engineering.

In sum, technology—or be it engineering—has contributed to man that small and sufficient advantage over the competitively hostile forces of nature that has enabled him thus far to survive and flourish. It has afforded him a measure of security, a slowly improving economy, and—most precious of all—moments of leisure, without which there can be little thought or reflection.

Modern engineering draws increasingly upon the methods and ideas of modern science and, in turn, contributes increas-ingly to scientific progress. Engineering is a partner with science and not a stepchild. Many of us have been aware and troubled by a curious mood too prevalent in some intellectual circles that arrogates an inferior grade to the part played in society by the engineer. I feel obligated on this occasion to reaffirm the impor-tance of his role.

The institution that you honor today was founded upon the idea that there is both worth and dignity in useful knowledge. The motto *"Mens et Manus"*—the mind and the hand—is in-scribed upon our seal. Nothing that one can foresee for the future diminishes the importance of a steadily advancing technology. I speak now not only in tribute to the engineer but on behalf of all those men and women whose efforts are dedicated to the translation of knowledge and ideas into products and services contributing to the health, nourishment, social stability, and economic security of mankind. I include in their number the doctor, the teacher, the architect, the manager of an industrial enterprise. All such professions have this in common—they combine art with science. In a limited sense, art is defined as the knowledge gained by skill, experience, and practice; but it is also, and in larger part, an intuitive feeling for the special meaning of the materials worked with and for the particular order that will best express this meaning. One might, then, say that engineering is art governed by reason or, equally, that it is reason organized by art. It is precisely this fusion of science with art that gives to engineering its special character and appeal. The engineer himself must take the highest view of his profession. The greatest opportunities and the most difficult problems lie directly before him— problems that challenge his originality, his intuition, his knowledge of the methods and data of science, his capacity to reduce ideas to practice, and, increasingly, his ability to manage the systems he has created.

The physical character of our cities and the quality of their design, the effectiveness of our future modes of transportation, the conservation of our natural resources, the synthesis and application of new materials, the creation of new industries, and the aid which we extend to other lands abroad, all these will be in large measure the work of engineers. These are vast responsibilities, calling for men of stature, and holding forth the possibility of great rewards in pride of achievement. They are worthy of the efforts of the most gifted young men and women of this oncoming generation.

I have spoken of science as the unremitting search for understanding and of engineering as its working partner. But men must be moved by more than an aimless urge to investigate and to fabricate. We have in our hands even now an almost limitless capacity to alter at will the material conditions of our existence. One can hardly conceive a technological goal that will not yield to the engineer, if men will but put their minds and wills to the task.

Yet, in this almost infinite array of possible projects, where shall we concentrate our efforts? It is all very well to assert that the applied sciences—medicine, engineering, and the rest—are directed toward useful purposes and the welfare of mankind. But such declarations are meaningless apart from a larger framework of judgment. Who shall say in the longer view what is useful and what is not? By what criteria and by what plan shall we make these great and often fearful decisions?

The simplest of the considerations involved is merely that of economy of effort. The limitations upon the human resources of a country, even of one so wealthy as the United States, are such that we cannot undertake all things at one time. And to this dilemma of priority, the advance of science now adds moral and ethical questions of a totally new order.

One may, to take an elementary example, spray the fields and the forests with insecticide. The mosquitoes die, as was intended. But so also, perhaps, do the birds and the insects, until over a vast area the delicate balance between fauna and flora is disastrously unsettled.

The advance of medicine enables the doctor today to conserve and to prolong life but without assurance that this life will be meaningful.

By the use of chemicals and radiation the biologist can induce mutations in genes and chromosomes, and so by an appropriate handling of genetic materials create new strains of living organisms. The same genetic laws govern the evolution of human beings. Conceivably someone may have the arrogance ultimately to undertake the breeding of a strain of "good men." But who then shall determine the new model of the "good" man?

In a quite different domain, as yesterday's Panel on Arms

Control warned, the consequences of a nuclear disaster might affect life on this planet for generations to come. It is just this inescapable fact that gives to the international problem of arms control a seriousness and an urgency without precedent in human history.

There is nothing in the new order of science that relieves any one of us of our personal, individual, moral, and ethical responsibilities. But science now presents to us, not alone as individuals but as societies and nations, moral challenges of a wholly new order of magnitude. The nature of these challenges must first be understood and then acted upon with all our energies.

Science, in sum, gives us knowledge and power of action. It tells us what we *can* do; we must turn elsewhere to learn what we *ought* to do. There is no certitude in man's affairs, and we learn by trial and error; but the errors are becoming increasingly expensive. For guidance we must turn to the accumulated record of all human experience, to the ethical teachings of our religious faiths, to the understanding revealed by systematic study of human behavior. These are the strands that through education must be interwoven with science and technology to form the fabric of a single culture and the hope of a harmonious and peaceful world.

And so tomorrow M.I.T. sets forth upon its second century dedicated to truth through science, proud of its concern for useful knowledge, and alive to a new order of ethical and social responsibilities.

9

Academic Freedom and Integrity

This speech was prepared for delivery at the Commencement at Johns Hopkins University on June 12, 1962. However, a sudden downpour of rain prevented Dr. Stratton from speaking.

As you go out from here today, whether to graduate school or to enter upon your active professional careers, you assume new responsibilities and some major obligations.

There is first your professional responsibility, your commitment to the world's work: to heal the sick, to teach, to develop and expand our economic and industrial resources, to seek understanding through science, to encourage and cultivate the arts, and to pursue these careers with an integrity and a constant striving for excellence that is the hallmark of the true professional estate.

This, however, is not all. There remain the overriding obligations of citizenship and your duty to support and defend the free institutions that are central to the idea of a democratic society. Pre-eminent among these free institutions stands the university. And because of your proper concern as educated men and women, I feel it appropriate to speak of certain dangers that constantly threaten the freedom of the American university.

First, I should like to impress upon you the importance of diversity in our national system of education and the urgent need to preserve it. That system is indigenous to this continent and throughout the whole history of the republic has reflected the variety of convictions, aspirations, and needs of the American people.

To a visitor from abroad it appears as a bewildering maze of public and private schools and colleges. It is a system which currently includes some 700 undergraduate liberal arts colleges, supported wholly by tuition and private philanthropy. To these we add the great, independently endowed universities. There are then the state universities, a number with enrollments of 20,000 or more—and some currently expanding into state systems of numerous campuses. There are also church-related institutions and an increasing number of junior colleges. And finally, there are institutes of technology—for the most part privately endowed —a number now evolving into scientific universities.

Again and again we read and hear that in this diversity lies the strength of American education; yet I fear that we take it all too much for granted. For the very essence of the democratic ideal is a belief in the worth of the individual—in his freedom, in his right to develop his full potential. And for all its obvious shortcomings, this seemingly chaotic system of ours is marvelously adapted to a wide range of human talents, interests, and capacity for achievement.

This same diversity of background, of views and attitudes toward the social, political, and intellectual problems of our time is carried from the college into our whole national life. In this very diversity lies our safeguard against a monolithic cast of mind and governmental structure within which neither free enterprise nor a democracy in our sense can survive.

The forces that currently are working against diversity and toward a national uniformity in higher education are neither conspiratorial nor of evil intention. They express in a way a kind of social entropy. In part they are the inevitable consequence of the growing size of our institutions and the need to plan and organize systematically. They reflect also in part a laudable de-sire to elevate standards. Yet for all this, every school, every college, and every university should retain its own identity and not attempt to copy another's. Its standards should be superlative, indeed, but they should be applied to its own aims and mission.

The demand for higher education is rising rapidly, both because of the expanding population and because of basic changes within our society. The task of meeting these demands necessarily and properly will fall increasingly upon the tax-supported institutions. But this movement places an even greater responsibility upon the independent colleges and universities. They have a role to play that is wholly out of proportion to their size. They must be the leaven in our academic life. They must be pace setters, innovators, and experimenters.

They, too, are subject to pressures toward uniformity. Although we in the United States have never succeeded in establishing common traffic laws, we are well on the way toward a single plan for college admissions. The College Entrance Examination Board presents us with a most puzzling dilemma: whereas these examinations have become absolutely indispensable in measuring one kind of academic aptitude, yet the system of objective tests imposes a rigid—and I believe limited—evaluation of the talents and potential of those selected for college.

The future of these independent institutions is by no means secure. The rising costs of teaching and research, the proliferating new fields of knowledge, and the dwindling resources of endowment are working against them. If they are to survive, we must support them.

I should like next to touch on that often maligned and widely misunderstood subject of academic freedom. There is a belief—rather common in some university circles—that long ago, perhaps in the twelfth century, there flourished an academic society in an original, Rousseau-like state of virtue, where freedom of thought and expression prevailed without restraint. They would have us also believe that the intervening centuries, particularly our own, have eroded away much of this freedom. This is a highly inaccurate reading of history. Even in that fabulous twelfth century, the ideas of such a towering figure as Abelard were condemned. Aristotle was proscribed for one generation only to be established as an ultimate authority for the next.

With the Reformation and the rise of nationalism, there sprang up all over Europe countless colleges and universities. None of these was conceived wholly in the idea of freedom of learning and teaching. Almost without exception each was initially established as an instrument for the transmittal of a particular set of beliefs, and each was severely limited by religious and political restraints. This was generally true of the first colleges established in the American Colonies. The early academic history of the United States is a record of never-ending inquiries into the orthodoxy of the faculty.

Then in the nineteenth century the supreme test of the principle of academic freedom arose with the controversy over evolution. For anyone of our own generation, it is difficult to comprehend the violent passions that were aroused by that issue —the bitterness, the reprisals taken against presidents of colleges, as well as professors, for a failure to conform to an established view. The battle was at a white heat in the year that Johns Hopkins was founded. Yet Gilman in his letter accepting the presidency spoke out eloquently for freedom of inquiry. He appointed an evolutionist as the first professor of biology, and he had the temerity to invite Thomas Huxley to make an opening address. These early events set the tone and the style of this university and made it a model for others to emulate. You may well be proud of the record.

Despite the contrary belief of many of my friends and colleagues, and notwithstanding the bitter recollections of the McCarthy era, it is my sincere impression that the faculties of our American universities have never enjoyed a greater latitude of intellectual freedom than they do today. But this is a judgment relative only to the past. The search for truth has no ending. The need to seek truth for its own sake must constantly be defended.

Again and again we shall have to insist upon the right to express unorthodox views reached through honest and competent study.

Today the physical sciences offer safe ground for speculation. We appear to have made our peace with biology, even with the rather appalling implications of modern genetics. Now it is the social sciences—the soft sciences, as the saying goes—that have entered the arena. These are young sciences, and they are difficult. But the issues involved—the positions taken with respect to such matters as economic growth, the tax structure, deficit financing, the laws affecting labor and management, automation, social welfare, or foreign aid—are of enormous consequence to all the people of this country. If the critics of our universities feel strongly on these questions, it is because rightly or wrongly they have identified particular solutions uniquely with the future prosperity of our democracy. All else must then be heresy.

There is a very simple way for a university to avoid controversy, to maintain the semblance of academic freedom. We need only to retreat within the ivory tower, to deal exclusively with the past or the esoteric, to reject all that is most relevant and vital to the life of our own time. No university worthy of the name will choose that course.

The movement toward the social sciences is illustrated very clearly by the experience of my own institution. M.I.T. for the better part of a century confined its program to the hard, clear facts of engineering and physical science. Today we believe that the problems generated by the impact of science upon society are no less urgent than the advancement of science itself. Consequently we, too, are moving vigorously into new fields of economics, political science, psychology, international studies, and industrial management.

I am sure that in these areas the years will bring both dissent and controversy; yet Johns Hopkins and M.I.T., and the other great academic institutions of this country, would fall short of

78

their mission should they fail to come to grips with the full range of difficult, vexing problems that are the product of our age.

We ask for freedom of study and protection for the rights of those who hold unpopular views. But there is no freedom without responsibilities and obligations. The public rightfully holds the university responsible for the intellectual integrity and professional competence of its faculty. And I believe that in areas of uncertainty and dissent, the university has an obligation to its students to present fairly and clearly the variety of views that give body and substance to the problem.

I come now finally to one further matter that affects the welfare of our universities—one that in its own way relates also to the problem of academic freedom. This is the subject of government support for basic research.

It was the Second World War that awoke Congress and the American public to the new role of science. We owed our principal successes in that conflict to an array of weapons and equipment that were the direct product of brilliant and very recent advances in science and engineering. The atomic bomb—the most spectacular as well as the most appalling—was a demonstration that no one could misunderstand. And it soon became equally apparent that the physical health and the economic progress of the nation, quite as much as its military security, would henceforth be dependent on our position in the forefront of science.

Thereupon the promotion of basic research and the encouragement of graduate study became national goals. Immediately after the war, the Office of Naval Research, through a highly enlightened system of contracts, began to make funds available to universities. The Army and the Air Force initiated programs of their own. In 1952 Congress established the National Science Foundation with its initial budget of $3.5 million. In the intervening decade this has grown to $261 million, and there must now be added equally substantial funds from other major federal agencies for the support of basic research.

The benefits of this policy to the country have been enormous. Not only has the volume of effort risen by orders of magnitude, but American biologists, chemists, physicists, and mathematicians have in these postwar years made impressive contributions to the quality of pure science. The federal government has invested heavily in science, and invested wisely.

Now the primary responsibility for effecting this tremendous amplification of effort has been assumed—and rightfully so, I believe—by our universities. And because of this concentration on science, and because of the very massive nature of support for research, the impact upon major institutions of higher learning has been profound. Without any doubt, the mood, the climate for intellectual life in the American university is in a process of change. We are, it seems to me, mid-course in a new phase of academic development. The immediate cause is a more abundant flow of funds for research. The larger cause is bound up with the powerful forces of change that are inherent in the scientific-technological revolution of our age.

Earlier I spoke of the role of the independent institution in our democratic society and the need to safeguard its freedom of action. Yet, given the nature of modern science and the national need, I see no alternative to a high and continuing level of federal support for basic research and graduate study, by no means to the exclusion of private funds but as a major supplement. Consequently, I think it of the utmost urgency that Congress, the major agencies of government, and the universities multiply their efforts jointly to develop wiser and more stable policies for grants and contracts than we now enjoy.

Happily there has been progress. The President's Science Advisory Committee, the National Science Foundation, and the National Institutes of Health have made a concerted effort toward this objective. Nevertheless, one need only read recent testimony before the House Appropriations Committee to learn that there are still many deeply imbedded misapprehensions on the relative functions and responsibilities of government and the universities for the advancement of science.

No federal grant for research should ever be viewed as a form of public philanthropy. The government indeed has only *one* justification for these grants: they are the means by which it fulfills its *own* responsibility to expand constantly the frontiers of science and engineering. Government turns to the university as the appropriate agent for the fulfillment of that purpose, and government owes the university in turn a proper concern for its stability and long-term welfare—for that, too, is in the public interest.

Policies that encourage the expansion of academic or research facilities, coupled with an unwillingness to provide adequately for the funding of the enlarged scale of operations, impose, at least upon private institutions, the necessity of drawing upon endowments, gifts, and other sources of current income to meet the deficit. Thus funds are diverted from the larger purposes of the university, from other equally important functions, and, indeed, from the salaries of its faculty.

But the dangers and the problems to be worked out are not simply fiscal, nor does the responsibility for their solution lie entirely on the side of government. It devolves upon the university itself to safeguard the balance between science and other branches of scholarship, to moderate the pressures and harassments of proliferating projects and reports, to maintain the prestige and the satisfactions of the teacher in an environment of research, and, at all costs, to preserve the intellectual climate of a university.

I began these remarks with some thoughts upon the professional estate. It is imperative to remember that society needs more than your technical competence; it needs, in addition, your willing and active dedication to the common good. Over and above the resources of intellect and training, we ask your commitment and support for the institutions of a free democracy.

The great movements of our time, the scientific-technological revolution that is transforming our entire society, are putting into new perspective the old tensions between freedom and control. Because of the traditions of Johns Hopkins, and not from any vested interest in the special prerogatives of the academic community, I have chosen to illustrate these thoughts by examples from the contemporary university. But the forces that are effecting change in our universities have an equal impact upon almost every American institution. Out of change we see emerging a growing complexity of organization, a drift toward centralization of control, an increase in size and scale.

These are powerful movements, but I am firmly confident that by the personal action and concern of educated men and women we may shape them to good ends. The preservation of individual worth, and of the quality of our special institutions, is still the highest goal.

10

Science and the Process of Management

A speech delivered at the International Management Conference in New York on September 18, 1963.

One of the greatest of novels—in the judgment of many, the greatest novel ever written—is Tolstoy's *War and Peace*. It is a story, as you may remember, woven out of events that took place in Russia during the Napoleonic invasions of 1805 and 1812. There is in that book something of everything that is human, and so one can hardly speak of a central theme; but over and over again the reader comes upon recurring scenes of battle to demonstrate the unmanageable waywardness and complexity of war. Great battles, says Tolstoy in effect, transcend man's power of information and decision. He speaks through his hero, Prince Andrew, who sums it all up with these questions:

> What theory and science is possible about a matter the conditions and circumstances of which are unknown and cannot be defined, especially when the strength of the acting forces cannot be ascertained? No one is able to foresee in what condition our or the enemy's armies will be in a day's time, and no one can gauge the force of this or that detachment. . . . What science can there be in a matter in which, as in all practical matters, nothing can be defined, and everything depends on innumerable conditions the significance of which is determined at a particular moment which arrives no one knows when?

There, I believe, is the clearest possible statement of the challenge with which we are concerned—a challenge not merely to the concepts of military command but to the very power of men to manage large affairs. It is not enough for the great leader to identify himself in some intuitive, mystical way with the will of the multitude. He must know the facts, the disposition of forces, and the true lines of battle.

Within the one hundred years that have elapsed since the publication of Tolstoy's classic, the progress of science and technology has added enormously to our capacity to observe, to record, to predict, and to control. Yet, despite the growing potential of these new tools of information, the problems to which they are addressed appear to be increasing at an even more rapid rate. And so one asks whether there are perhaps inherent limitations on the capacity of man to assimilate information, to sort out relevant facts within a finite time, and then through the power of intelligence to analyze and organize the data, to form judgments, and to communicate decisions. Can the function of leadership, traditionally vested in the single individual, be successfully expanded and multiplied by delegation and decentralization of power and responsibility? Or are there natural laws as yet undiscovered that govern and limit the ultimate scale of human effort? Is it possible, after all, that time may prove Tolstoy to have been right in his views of history and society? I know that any such thought runs completely counter to the spirit of our meeting today, and indeed I, too, take a much more optimistic view of the future. Yet I think we should be on our guard lest we be wholly captivated by the marvels of modern electronics.

The predominant social movement of this century is a constant increase in the size and complexity of all our institutions. It appears in the enlarging corporate structures of business and industry; it is obvious in the irreversible expansion of governmental agencies; and—if I may speak out of my own experience—it is creating major problems in the development and conduct of our universities. In every division, in every department, the numbers of people are growing. A change of scale is taking place everywhere in the organization and the sheer magnitude of operations. We observe it in a multiplication of administrative units, in an expanding network of interdependencies, and in the diversity of services and products.

In effect, we have been asked to consider today how far the tremendous outpouring in recent years of new devices and instrumentation, of high-speed computing machinery and communication systems will go to fortify management in its endeavor to impart order and intelligence to these vast complexes of people and operations. For the principal burden of seeking answers to a great array of questions, and of resolving such problems, will fall upon management itself. And as this process goes forward, judgments and decisions must be worked out in the context of an exceedingly sophisticated environment of science and engineering.

I might express all this in another way by saying that the basic information network of modern industry is the tightly coupled system of science, technology, and management. Progress and change in one domain immediately and materially affect the problems and potentials of the others. And so I want initially to comment upon some characteristic features of this system and the requirements they impose upon managerial leadership.

First of all, one must clearly take account of the role and the new prestige of basic research. I need not emphasize that one of the most notable sociological consequences of the Second World War was a sudden awakening of public interest in fundamental science. The war itself had demonstrated that the continuing progress of science had become indispensable for the maintenance of national security. The nature of the process of discovery, development, and application had been revealed in such dramatic fashion, and the successive steps had followed in such a short space of time, that no one could fail to grasp the wider import for furthering the industrial economy. Here in the United States the lesson was impressed not only upon the public but also upon Congress. For the first time substantial social and economic values were attached to mathematics, physics, chemistry, and biology, and their advancement became a national goal. Our federal government, through its manifold agencies, undertook the support of fundamental research on a massive scale. A significant part of this support has been allocated to universities in the form of grants and subsidies, with far-reaching implications for the character as well as the administration of our institutions of higher learning.

But universities are no longer the only contributors to basic science, for the potential of research has not been lost upon industrial management. By every measure this new involvement constitutes a major change. I might point out that twenty years ago the contents of the *Physical Review*—the outstanding American journal of pure physics—consisted almost exclusively of scientific papers from our universities. A survey of last year's issues reveals that only 50 per cent of the contributors were affiliated with academic institutions. The balance of the papers originated in the growing multitude of industrial and governmental laboratories.

The yield from this massive effort—national and world-wide— has been tremendous. It has been costly indeed. No doubt, too, it has on occasion been inefficient. Yet the fact remains that the frontiers of science have been pushed forward with unprecedented speed. We see it in the expansion of existing areas of knowledge and, even more strikingly, in the emergence of entirely new fields. But the most significant of these developments from the point of view of our present discussion is the rapidity with which these scientific discoveries are taken over into engineering and converted into useful products. The time gap between the discovery of the physicist and the design of a commercial product is rapidly closing—closing, in fact, so completely that the process is now essentially continuous.

The economic implications of this foreshortening of the time scale are clearly apparent. They bear particularly upon the constant, intensive search for new products and improvement of the old. They are reflected in the establishment by the basic industries of an increasing number of purely scientific laboratories and the encouragement of close ties to the universities. They underlie the host of new ventures that have sprung up to capitalize on the advances of recent research. They are affecting our whole philosophy with respect to basic patents. And, finally, they are bringing about an increase in the pressures on management to accelerate decisions that affect future development.

Now, in this process of transfer from discovery to consumer, engineering is the bridge between science and management. And as I look back over the developments of recent years and reflect upon the changes that have overtaken the engineering profession —the kinds of problems with which the engineer now has to cope, the reorientation of his views imposed by the advance of science, the new emphasis upon research—I find a striking parallel in the current evolution of management.

There is, of course, nothing new in the idea that engineering draws heavily upon the physical sciences, any more than that management must rest upon a firm foundation in economics. But lately pure science has been spilling over into technology, and the basic scientific content of engineering is steadily on the increase. It affects not only the materials with which the engineer has to work but, more significantly, his methods of attack, his whole approach to the problems of development and design and, above all, it imposes a new order of demands upon his technical training.

Since the Second World War the progress of these changes has been so swift that we have had here in fact a revolution. For a profession as old and staid as engineering, adjustment to this new state of affairs has by no means been easy, and the stress and strain of accommodation has sometimes been painful. We in the technical universities feel it most acutely in the conflict of views on engineering education.

The idea of the engineer as a man of practical action is as

ancient as written history. It is a tradition handed down from those men of Caesar's legions who first pushed bridges across the Rhine. And it remains today in the need for men with competence and experience to construct our roads, our bridges, and the great buildings of our cities, and to design and maintain the operations of our basic industries.

But rising among them is this new breed of scientist-engineers, brought up in research, whose point of view seems more academic than practical, and yet who clearly hold the key to the future of industrial development. It is to such men as these that we owe much of the modern electronic industry, the new uses of metals and plastics, and essentially the whole of information technology.

Modern management, considered as a profession, offers, I believe, a complete counterpart to these trends. Management, of course, is many things—a vast domain that encompasses diverse occupations at various levels of sophistication. It includes the specialists in finance and labor, in sales and distribution, in accounting and production. It has its own traditions of pragmatic judgment and decisive action. More, perhaps, than any other profession, it represents a synthesis of all the elements of society, economic, political, and technical.

But now the ideas, the products, and the methods of science and engineering are invading the field of management, and management, in turn, by necessity will penetrate more directly and more deeply into the technological heart of modern industrial operations. The research laboratories of which I spoke earlier are but one example. They are a responsibility of management, and if they are to be in fact effective instruments of growth, there must be on the part of management more than a superficial understanding of the nature of research, of the motivation of the people who engage in such activities, of the natural environment in which research thrives, and of the process of transfer from laboratory to useful product.

There are currently many efforts, as you know, to establish vigorous scientific approaches to the problems of management.

Most of these draw upon the mathematical methods of probability theory, statistics, and operations analysis. The same trends can be observed in the underlying social sciences—particularly in the fields of economics and psychology. The key to this progress is an increasing capacity to make meaningful measurements, and so to reduce theory to quantitative results.

Let me touch upon one other development. Of all the concepts that have flowed out of the recent progress of engineering, the idea of a system as a total entity has by far the most meaning for management.

A generation ago the major efforts of industrial engineering were focused on the design and manufacture of components. I am speaking, of course, of the vacuum tubes and coils and condensers of radio sets, the pistons and wheels and radiators of automobiles. The assembly of these separate components into total products presented no additional technical problems of any great magnitude. But as our technology advances, as the complexity of these assemblies of components grows, the prediction of how these several parts will interact, of how the ultimate system as a whole will behave, emerges as the truly difficult problem of design.

Some years ago we in the United States began to interconnect the electrical transmission lines of many utility companies into great networks. That was perhaps our first introduction to a major systems problem. With the methods then available it was impossible to foretell the effect that a stroke of lightning, a short circuit, or the failure of a generator at one point might have at another.

To take another example, an airplane is an assembly of components, an intercoupling of engines, wings, and ailerons. The central and most difficult problem of aeronautical design today is that of predetermining the behavior of the aircraft in flight in response to a shock or impact at one or another point of the system.

This is no longer a task for the handbook engineer. The answers to such questions about stability and response seldom can be derived out of rough calculations or an intuition that is the product of past experience. Systems analysis represents the most intricate phase of modern engineering. To draw upon the terminology of another profession, one might say that we have graduated from the anatomy of components to the physiology of an organic whole.

Now the concept of an assembly of parts and operations whose integrated behavior is subject to analysis and design has led us to an immediate and vast extension. For in fact the whole chain of industrial processes that begins with research, that passes into development, design, and then on into production, distribution, and sales, constitutes a tightly coupled system. It is a chain that closes upon itself, a system for which we can construct electrical or mechanical analogs exhibiting comparable properties of stability, feedback, and response, and governed by a new industrial dynamics. And so we see here engineering drawn into the processes of management, and management, for its part, entering earlier and ever more deeply into the mainstream of technology.

The basic premise for our whole discussion is, of course, that science and engineering constitute the underlying matrix of change in all our contemporary institutions—social, industrial, academic, and governmental. In my summary thus far I have wanted merely to identify three elements in that matrix which are contributing, most powerfully, I believe, to change and the growing complexity: research as the driving force of innovation, the rapid transfer from discovery to useful application, and the new concept of the engineering-managerial system.

Let me turn now more specifically to this "information revolution" and to the questions that I posed at the outset of my remarks.

For the engineer, this term *information* has a precise but technical definition. It derives from a mathematical theory which analyzes the basic nature of the information contained in a message and of the process of communication between men and men or between men and machines. For the first time it has provided us with a quantitative measure of informational content, based on answers to a set of questions that admit only the elementary decision: yes or no. Thence one may proceed to develop the whole mathematical structure that underlies all systems of communication, computation, prediction, and control. The same principles and the identical logic apply to mechanical and electrical systems, to physiological networks comprising muscles, nerves, and brain, and to the syntax and grammar of language.

The development of the modern high-speed computer, the multitude of devices encompassed in the rapid development of automation, are *technically* possible today thanks to extraordinary engineering advances in the field of electronics. But the conception of these systems, and their functional and logical design, are the practical consequences of abstract information theory. Indeed, we have here a striking illustration of the interdependence and interweaving of pure science and hardheaded engineering. Information theory has within the past few years been the most powerful stimulus for electronic development; and the electronic industry, in turn, has made that theory practically as well as intellectually fruitful.

Out of these advances has come the change of scale, the enormous magnification in our capacity to record the results of unit operations, to feed that information back to a central, focal point, to reduce the mass of data to intelligible form, and to retransmit the ensuing decisions to distant operating points for instruction and control.

The technological horizons of this new field of information now appear almost unlimited. This is the domain of the engineer, and we can count upon him to push forward, to enlarge the storage capacity of computers, to increase flexibility and speed of retrieval.

Yet this wonderful technological triumph of our age, this abundant harvest of new and faster devices for processing data, will not alone resolve the problems of complexity. Without a clear understanding of their limitations, we may risk piling one order of complexity upon another. Once set going, our information system has the capacity to drown us in a flood of numbers reminiscent of the old legend of the sorcerer's apprentice.

The desperate predicament of the modern library serves in a way to illustrate the point, for the library is a focal center for the vast outpouring of papers, journals, and books which is one of the marks of our time. Some day we may design a computer to search this huge storage of information with a flick of a switch. And what will be the result? A contribution to learning, or merely an enlargement of the index? And there you have the essence of my thought: learning and information are not equivalent, and an informational system is valuable only as it adds to knowledge and understanding.

In the context of these ideas technology shows what can be done and provides the means; managerial leadership in the highest sense must tell what needs to be done. It must look to the larger goals, set the objectives, and thereby give form and meaning to the search for information.

Were it not for the fact that management is indeed assuming this responsibility and responding to this challenge, the tremendous expansion of our corporate structures would already be leading us into an organizational jungle. Fortunately, the information revolution is generating more than hardware and a flood of data. The extraordinary technical gains have been matched by a corresponding increase in our capacity to integrate the functional areas of management and observe the factors that influence growth and stability. They have opened new and penetrating insights into the nature of the managerial function itself. They are stimulating fresh ideas about the structure of organizations, about how policies are formed and decisions made, and about the decentralization of authority and responsibility.

I have commented earlier upon the necessity for management to penetrate deeply into the technological heart of industry. I am not suggesting for a moment that we should create managers in the image of scientists and engineers; but if we are to capitalize fully on our technical progress, then management as a whole must develop a scientific competence to move freely in an environment of science and technology. Just as the progress of science has irreversibly affected the profession of engineering, so science and engineering both are placing their imprint upon management.

Yet, for all my personal faith in the potential of technology, I am far from believing that the machine is about to replace man. It can do chores for the mind at a speed and on a scale far beyond human ability. But a machine can teach us no ethical principles. It has no set of social values. It can contribute nothing to the spirit and to direction according to rightful purposes. These are uniquely characteristic of men, who through supreme qualities of leadership express judgments, beliefs, ideals, and views that involve more than information.

Modern man is the child of technology. It is influencing and shaping the progress of all his affairs. But though we may be the children of technology, we must be its masters and not its slaves. The whole objective of our new and increasing technological power must be to improve the human condition and to help fulfill the aspirations of man. And this, too, I submit, is the highest purpose of management.

11

Learning and Action

The Penrose Memorial Lecture delivered at the spring meeting, on April 23, 1964, of the American Philosophical Society, Philadelphia, Pennsylvania.

As I look back upon the names of those who have delivered the Penrose Lecture in other years, I am conscious of the great honor that has been paid me by your invitation to speak here this evening.

The Penrose lectures of the past have dealt with a most extraordinary diversity of subjects. One lecturer has told you that the earth is a clock, and another has suggested that the universe was running down. You have heard about the gene, about Lewis and Clark, and about prospects for space travel. Last spring René Dubos gave a very beautiful paper on "Logic and Choice in Science."

There was a time when I might have ventured to discuss the implications of my own field of physics. But college presidents these days soon forswear their scholarly competence—at least if they have been physical scientists.

And so I shall turn to the subject that has come to occupy much of my thoughts—the university as a social institution. There are innumerable forces acting to modify the character, the academic processes, the goals, and certainly the outward appearance of our universities. I shall endeavor to describe tonight only a single aspect of this shifting, changing scene.

Let me begin with a bit of history.

On the morning of October 17, 1940, several members of a small committee under the leadership of Alfred Loomis met for breakfast with Karl Compton at the Wardman Park Hotel in Washington. Dr. Compton came to that meeting in a dual role: as President of the Massachusetts Institute of Technology and as Chairman of a major division of the National Defense Research Committee (NDRC). His Division D was charged with the task of applying physics and engineering to the instrumentalities of war. Pearl Harbor was still more than a year away; but France had fallen, and although the Germans had lost the first round of the Battle of Britain, the outlook for that coming winter was desperate. In September the Tizard mission had arrived to disclose in secret the development of a magnetron for generating very short radio waves and to ask for American help. Their most urgent and critical need was a lightweight, short-range radar for the use of night fighters operating over London. The new magnetron held out this possibility and could be the key to survival. Only in the United States, with the wealth of our technical resources and freedom from constant bombing, was there a hope of producing in time an effective piece of equipment. The newly organized NDRC had responded to the appeal; and the decision to be made on that October morning was to fix upon a location for the project.

Academics were in a minority among the members of Dr. Loomis' Microwave Committee, but there was powerful representation from the leaders of the electronics industry—from the Bell Laboratories, General Electric, Radio Corporation of America (RCA), Westinghouse, and Sperry. The Committee thought first of placing the new project in the neighborhood of Washington and under military administration. The pioneer American work on radar had begun at the Naval Research Laboratory, and Bolling Field was nearby. But there were doubts whether the Army or Navy had in fact a serious interest in centimeter wave lengths; and there were also doubts on the prospects for their effective collaboration with civilian scientists. The Committee next considered industry, but fared no better in its efforts to settle upon an appropriate contractor. Only then, after weighing the relative merits of both government and industry, was the proposal made to turn to a university. In the end M.I.T. was asked to undertake the contractual responsibility for the project. After much urging, Dr. Compton acceded to the request, but reluctantly and with grave misgivings.

So it was that the Radiation Laboratory came to the M.I.T. campus. The name was a cover, chosen intentionally to confuse its real purpose with that of a very different laboratory in Berkeley. Over the succeeding five years it was destined to draw physicists, mathematicians, chemists, biologists, engineers, and even architects and historians from every part of the country, to grow in size to some 4,000 people, to expend more than 100 million dollars on the invention and development of microwave devices.

Promptly with the end of the war in August 1945, the Radiation Laboratory in Cambridge was terminated. The men and women who had been its life and energy dispersed to their own universities, to industry, to their private affairs. A few months later there appeared a brief, informal history, and there, in an assessment of five intensive years, one may read this statement:

> Perhaps the deepest influence of the Laboratory could be found in its beginnings. Here a new dynamic declared itself. Before this, science was a random spore: at the mercy of a thousand accidents, or irrational controls, and most of all of poverty. And even now during the war the area was circumscribed, the broad aims laid down. But still, in this area, the men of science were given great liberty; and before they were through it was evident that they had created a new form. This was the first appearance, in the national fabric, of science as an autonomous force, and of the scientific method as a way of work. Here for the first time in significant amounts scientific energy was set free.

In the perspective of history those words may now seem a little exaggerated, even slightly pretentious. Because of the continuing veil of secrecy the authors knew little of the details of another great project brought to fulfillment on a high mesa of New Mexico. Nevertheless, in essence, and as an expression not of any single project but of a national effort, these lines were prophetic. The years 1940 to 1945 represent a great watershed in the evolving relations of government and science. In the postwar era the cultivation of science was to become a national goal. From 1940 to 1963 the support of research and development through federal funds alone was to increase some two hundredfold. The impact of this new force has made its mark upon all the institutions of our society: upon education, upon industry and the organization of our economy, upon the agencies of government, and even upon the conduct of foreign affairs.

To the academic world it has brought both blessings and problems. Thanks to the new policies of support, wisely administered by the National Science Foundation, the National Institutes of Health, the Department of Defense, and other major agencies of government, science on every frontier has pushed forward at a breath-taking speed. Great centers of research have emerged, and with the influx of government funds, many of our institutions—both public and private—have begun to take the form of "Federal Grant Universities" that Clark Kerr described so brilliantly a year ago in his Godkin Lectures.

Yet at the same time we are beset by many anxieties. There is a widely prevalent concern for the erosion of traditional academic values. There is concern, too, for the emphasis upon the physical and biological sciences—the "hard" sciences—at the expense of the humanities and social sciences, for the emphasis upon research productivity at the expense of the undergraduate, and for the concentration of talent and resources in too few institutions. There are apprehensions about the very bigness of modern science and the suppression of the individual creative personality by the massive force of the team and center. And there are fears that the machinery of grants and contracts must inevitably shape

97

research and scholarship toward popular objectives, leading in the end to an encroachment of federal control upon the intellectual freedom and integrity of the university.

These questions have been the subject of countless papers, lectures, and editorials, and of studies by the President's Science Advisory Committee, by the Association of American Universities, by the National Academy of Sciences, and recently by various Congressional committees. There is no need for me to dwell upon either the nature of such problems or their importance; for all of you here this evening have in one way or another had a part in this vast revolution of our own age. Yet there remains one aspect of this great complex of growth and change that deserves our most serious attention—the growing involvement of many of our larger universities in certain kinds of nonacademic enterprises.

A university is a social institution designed to serve three roles:

First, to convey to each oncoming generation the accumulated learning of the past and to cultivate taste and style;

Second, to generate new knowledge and so to expand the horizons of the human intellect;

And sometimes third, to serve the community and nation directly through its faculty and through the use of its material and administrative resources.

Historically all of these functions have been fulfilled at one time or another in varying proportions. What is new is the scale with which research and public service have now emerged upon the scene, with all the consequent issues of balance and moral obligation.

I began my remarks with a story about a certain breakfast-table conference. The point of that story was that it involved a choice and a crucial decision—a decision to entrust to an academic institution contractual responsibility for a project whose objectives were anything but academic. The occasion has seemed to me worth marking, for in a way it set the pattern for a series of actions to follow and a wholly new outlook upon the uses of the university.

Let us go back for a moment in time.

The dialogue on the two purposes of learning—the pursuit of knowledge for its own sake versus knowledge in the service of mankind—must be nearly as old as learning itself. Eric Ashby has quoted an epigram from Epictetus in support of basic research: "though sheep eat grass, it is wool which grows upon their backs." The great universities of the twelfth and thirteenth centuries, which we tend to look upon as models of original academic virtue, were in a way vocational schools, clustered about a faculty of arts and preparing for medicine, law, and theology—the most respected occupations of that day. There were the Encyclopedists and the Enlightenment of the eighteenth century, followed by the British utilitarians in the nineteenth. If I remember rightly, Friedrich Engels—not my favorite authority—maintained that learning has merit only in so far as it leads to action. And to all these purveyors of utility—to Bentham, Mill, Huxley, and the rest, as well as to Bacon—Cardinal Newman responded with his eloquent, classic, wholly unrealistic essay on "The Idea of a University."

Both traditions played their part in the formative years of American higher education. The older liberal arts colleges reflected the influence of Oxford and Cambridge. Jefferson and Franklin, of course, spoke for the usefulness of knowledge, and a number of institutions in the first part of the nineteenth century, including among them New York University in 1831, were founded upon this idea.

But we must look to the land-grant colleges for the first great innovation in higher education upon this continent. They were truly indigenous to our soil, an authentic American contribution. In a wholly new form they expressed the importance of useful knowledge, and the idea, as a president of Ohio State University once declared, that "an institution is to be operated for the good it can do; for the people it can serve; for the science it can promote; for the civilization it can advance."

Under the Morrill Act of 1862 these colleges undertook the education of men and women destined principally for agricultural occupations and the mechanical arts. Everyone is familiar with the story of how many such colleges grew into great state universities and how they have contributed to the American ideal of an open, mobile society. But I should like to recall one unique feature of the early land-grant institutions that foreshadowed the developments of our own day. Experiment stations were directly affiliated with the colleges. Their purpose was the general improvement of the agricultural community. Their ties extended to the academic faculty, on the one hand, and to the county agent, the soils expert, the timber expert, and the drainage specialist, on the other. For the first time we find under the aegis of the university an organization whose chief interests were related only indirectly to education, and the first significant example of federal grants to academic institutions.

In the First World War, it appears that science, and certainly academic research, played a relatively small part. Ballistics was a problem for the military. Airplanes seemed barely practical and strictly for the engineers. The detection of submarines did present a new and challenging technical problem, and there the Navy turned primarily to industry for help. Probably the most notable exception was in the case of the Chemical Warfare Service, which drew substantially upon universities and academic scientists for assistance. Yet on the whole the participation of university scientists, even in association with the military and with industry, was modest by our present standards.

Dr. DuBridge has pointed out that

    . . . when President Wilson appointed a scientific board to advise the Navy Department, he made Thomas Edison its chairman. . . . There was only a single physicist on Edison's board, and he was put on because, as Mr. Edison said, "there ought to be one mathematical fellow around in case we need to calculate out something."

Institutionally the role of the university was limited almost entirely to that of education through the Student Army Training Corps and special educational programs, such as ground schools for pilots, signal corps schools for wireless operators, and intensive courses in naval architecture.

The National Academy of Sciences did endeavor in 1916 to focus the academic resources of the country by the formation of the National Research Council. The National Advisory Committee for Aeronautics was established "as an independent agency under a governing board that included a number of scientists from private institutions, serving the government only part time." Its first research project was done through a grant to a private institution—to the Massachusetts Institute of Technology, in the amount of $800. You see how times have changed!

Then in the sharpest contrast to everything that had gone before, science profoundly influenced the whole course of the Second World War. As I noted earlier, President Roosevelt, even before the onset of hostilities, had established the National Defense Research Committee. Its membership was dominated by men drawn directly from academic life—Bush, Conant, Tolman, Compton. Jewett from the Bell Laboratories represented the National Academy of Sciences. The role of invention was represented by Coe, the Commissioner of Patents. Liaison with the military was attained through General Strong for the Army and Admiral Bowen for the Navy. It was natural in the light of these circumstances that the NDRC should turn to the universities for help. Before the end of the war several hundred million dollars had been expended through universities for military research and development.

The NDRC—or perhaps I should speak of the Office of Scientific Research and Development, which included also the Committee for Medical Research—recognized the role of science and engineering in modern war, demonstrated how science may be mobilized for research, and in the process moved the universities into the arena of action on a scale and in a manner that has no precedent in academic history.

The most radical innovation was the idea of a massive attack on a complex problem through the establishment under university administration of a few great research centers for an organized scientific effort.

I have already spoken of the Radiation Laboratory at M.I.T. The origin of the Manhattan Project is even more familiar— indeed, I know that some of you in this audience played an active role in its beginnings. In the final days of January 1942, while fundamental experiments on fission were continuing at Columbia and Princeton, Arthur Compton organized the Metallurgical Laboratory at the University of Chicago. While Fermi was leading this development to its epochal success, Robert Oppenheimer suggested a special bomb laboratory and together with McMillan and General Groves chose the property of the Los Alamos Ranch School in New Mexico. The Army undertook supervision of the initial construction, but the task of organizing and recruiting the people who were to make the laboratory and to carry on its scientific work was entrusted to the University of California.

At the California Institute of Technology the work on rockets led to the establishment of the Jet Propulsion Laboratory. At Johns Hopkins University the Applied Physics Laboratory undertook to translate a variety of scientific ideas into devices for combat use, and became identified particularly with contributions to the development of the proximity fuse.

There were others, but these are sufficient to recall the circumstances and the pattern of these organizations which sprang up almost overnight on some of our campuses and which have left a permanent mark on the American academic scene.

All of us who had a part in the events of those stirring days thought of them solely as a wartime phenomenon that would end with the termination of hostilities. We looked forward eagerly to a return to a normal academic life in the manner of the thirties. And indeed, even before the armistice with Japan, the moves had begun to dissolve the NDRC, to terminate contracts with the universities, and to liquidate most of the larger projects.

Yet very shortly we were to discover that things were never to be the same again. The public had been awakened to the importance of science and had grasped its implications not only in the areas of defense but also in stimulating the economic advance of the nation. For the scientist himself the wartime developments brought an enormous change in the scale of his enterprises. Even those institutions that clung most tenaciously to the conviction that the search for knowledge for its own sake is the primary purpose of a university were confronted by a whole new order of dimension in the cost and organization of research. To the most individualistic scholars it became clear that there were areas in which progress could be made only by collaborative efforts and by large installations of equipment and laboratory apparatus. Ernest Lawrence had already demonstrated with his cyclotron at Berkeley the effectiveness of a group attack upon nuclear structure.

The postwar years saw not only further collaborative efforts among scientists, and on an expanded scale, but also a sharing on the part of universities in the organization and management of great research centers. The Brookhaven National Laboratory was one of the first, but the National Radio Astronomy Observatory at Greenbank, West Virginia, the National Center for Atmospheric Research at Boulder, Colorado, the Princeton-Pennsylvania Proton Accelerator, all are modeled essentially in the same pattern and spirit. Their purpose is the advancement of basic knowledge, and they are actually a projection of the traditional functions of the university, an adaptation to the technical imperatives of our age. They represent no real departure from the past but only a change in degree—a new phase in the continuing advance of science.

But the war gave birth also to a quite different type of establishment. In April 1946, the University of Chicago accepted a contract to operate the Argonne National Laboratories for reactor development on behalf of the Atomic Energy Commission. The University of California retained management responsibility for the Los Alamos Laboratory and added work in the field of atomic energy at Livermore. The Jet Propulsion Laboratory at California Institute of Technology was extended first under contract with the Army and later for the National Aeronautics and Space Administration (NASA) as one of the nation's principal centers for space research and development. Johns Hopkins continued to operate the Applied Physics Laboratory for the Navy.

The Korean War gave a further impulse to the growth of these enterprises. In 1951 the Lincoln Laboratory was formed at M.I.T. in an effort to bring developments in the field of high-speed digital computers to bear on the problems of the air defense of the North American continent.

Each of these centers and a number of others like them have been built upon the advancement of some field of science and engineering, and they have, in fact, contributed in a major way to the progress of knowledge. Yet they differ fundamentally from the older modes of academic organization in that their very existence has been predicated on the fulfillment of some national purpose.

As one looks back upon how this whole new type of enterprise came into being, it is clear that the successful experience of the war and a normal desire to adhere to proven models played a very important part. The Cold War, the climate of continuing emergency, the sense of national obligation had a profound effect on the attitudes of many university administrations. But the real cause, I believe, lies deeper. From the new science—and, even more, from the new technology—have emerged problems and projects of a wholly new dimension and complexity, demanding for their solution the collaborative efforts of a wide range of specialists. And it is precisely in the availability of such diverse specialists and in their experience in welding together groups of scientists for a massive attack upon a particular problem that the universities seem to offer conspicuous advantages.

There were of course, from the outset, efforts to find other solutions. Thus in 1946 the Air Force, appreciating keenly the need to continue fundamental research on military weapons and the conditions of their employment, initiated the RAND Project under contract with the Douglas Aircraft Corporation. This was a pioneering effort. Douglas anticipated that a consortium among the aircraft manufacturers could be formed to sponsor the work. That hope for industrial cooperation soon proved futile because of issues involving conflict of interest and competition for contracts, whereupon Douglas divested itself of the project and recommended the independent establishment of the RAND Corporation, one of the first of the new breed of nonprofits.

When it became clear to both the Atomic Energy Commission and the University of California that the laboratory at Los Alamos should be freed from a major involvement in nuclear hardware engineering and production, the Commission turned to the American Telephone and Telegraph Company for help. The Company reluctantly accepted the task, with a concern akin to that of a university faced with the same problem. The new assignment would represent a diversion of its efforts into fields remote from its main interests. There were troublesome implications bearing upon its relations to government and other members of the industrial community whose support was essential to success. In the end the Company undertook the task by forming the Sandia Corporation as a nonprofit subsidiary, which in turn became still another organizational prototype.

In 1956 a situation of a different kind arose. The Weapons Systems Evaluation Group (WSEG) had been created by the Joint Chiefs of Staff to carry out operations research and to fulfill a function somewhat narrower than that of RAND but of the same general character. WSEG, however, had been established initially within the Defense Department under civil service regulations. Despite the efforts of highly competent research directors on loan from universities, experience had shown that it was impossible to assemble and maintain a well-qualified technical staff. The Secretary of Defense turned first to a single university, which declined to accept sole responsibility but did agree, ultimately, to form a consortium of universities for the purpose. The result was the establishment of another nonprofit organization—the Institute for Defense Analyses—whose board of trustees represents eleven universities.

One can hardly say that the rise of these new enterprises has passed without notice. They have been attacked in principle from within the universities, and by industry as unfair competition. Various agencies of government and members of Congress have looked upon them as a threat to the civil service system and to the development of quality within the government establishment. The complaints of industry, quite understandably, are focused upon the intrusion of nonprofit institutions into what they have good reason to consider to be their own preserve and upon the competition they offer in the recruitment of the highest quality of scientific personnel. The advent of NASA, with its large space projects, has aggravated the situation, and anyone wishing to get a feeling for the bitterness that has grown in some quarters need only scan the letters published in a variety of trade and technical periodicals following the mishaps in February [1964] to the Jet Propulsion Laboratory's Ranger 6. The lack of substance and the disregard of the facts in many of the letters do not in any way lessen the reality of the tensions.

Two years ago criticisms from both within and without the government had reached such a pitch that the President established a committee under the chairmanship of David E. Bell, then Director of the Bureau of the Budget, and including the Secretary of Defense, the Administrator of NASA, the Chairmen of the Civil Service Commission and the Atomic Energy Commission, the Director of the National Science Foundation, and the Special Assistant to the President for Science and Technology. The Committee was charged with the examination of all aspects of the problem of government contracts for research and development. It dealt with the problems of conflict of interest and the erosion of competence in the government's own research and development establishments. It made a serious effort to define the individual and special roles of universities, nonprofit corporations, industry, and government laboratories in the fulfillment of national objectives. It acknowledged the criticism that the use of contracts with nonprofit organizations appears often to be a subterfuge to avoid the restrictions of civil service salary scales. Nevertheless, it concluded that universities generally, and some of the newer forms of nonprofit corporations, are indispensable for carrying out complex projects in the public interest, and offered remedial measures for strengthening the government civil service.

Admittedly I have presented only a very sketchy account of an amazingly intricate complex of laboratories, research centers, and corporations that have mushroomed in the United States within the past eighteen years and that differ radically, both in character and in the scale of operations, from anything we had previously known. These organizations differ among themselves in corporate form and administrative structure. Some are associated with an individual university. Some are sponsored by consortia of universities, and some, as in the case of RAND, simulate an academic environment but are wholly free of direct university ties. But all of them have one thing in common. They originated from an effort to master some problem or web of problems arising out of the new science and the new technology.

Laboratories or projects managed by universities have been undertaken, in almost every instance, because of a real and urgent need and in the absence of any apparent alternative solutions. I believe that this can be said, too, for the majority of the nonprofit corporations. One can easily demonstrate that, taken as a whole, these institutions have contributed enormously to the military security of the United States; they have been responsible for a number of outstanding advances in science and engineering over the past decade; and, thereby, they have vastly enhanced the productivity of industry itself. By virtue of the close coupling between engineering and science, the results of our progress have fed back into the university, revitalizing engineering education and stimulating basic science as well.

And yet for many it is difficult to reconcile the scale and character of these new developments with the historic mission of a university. Is there not something incongruous about the management of our national affairs that universities should be charged with so large a measure of responsibility for defense research? Should we not in the long-term interests of the country diligently seek to devise other means of fulfilling these particular obligations? No doubt we should.

But the fact remains that when one analyzes this revolution of science about which we speak so often, one observes that it is generating more than hardware and certainly more than defense technology. Out of it are emerging great problem areas and systems—highly technical, very complex, drawing upon scholarly resources that at the moment exist only in our universities but bearing directly upon the progress and welfare of our whole society. It may prove both wise and possible for the universities some day to divest themselves of projects that revolve principally around matters of national security. But I do not believe that they will ever again be able to isolate themselves wholly from enterprises that bridge the interests of the scientist and the scholar with the world of action.

The transformation of our cities is a direct consequence of technological change, often bringing with it the physical and social degradation of large areas. The crisis in transportation remains one of our most baffling and critical problems. The shift to automation in industry is accelerating and will have profound effects on the character of our labor force, upon its training, and upon its security. There are problems of public health that draw upon the social sciences as well as upon biology and medicine. The highly controversial use of pesticides relates questions of ecology, of conservation, and of the improvement of agriculture. The proper development and use of our water resources may have a larger influence upon the growth of the country than limitations on fuel. And then there is the progress of education itself: how are we to store the knowledge that we are accumulating, how can we retrieve it at will, how will we develop both the content and the methods of education to keep pace with the quickening advance?

These examples, although taken at random, exhibit certain common features. None of them falls within the domain of a traditional discipline; neither physics nor biology nor chemical engineering alone could contain them. All of them touch upon areas of economics, political science, law, and management. Some involve psychology and city planning. They must be attacked from several points of view simultaneously; and these many-pronged attacks must be wisely and effectively coordinated. Just as they extend beyond the framework of any one discipline, they are also broader than the terms of reference of any single agency of government. Nor is it obvious that any existing form of industrial enterprise could be expected to have at hand the variety of professional resources necessary for such undertakings. Moreover, although these are all problems of the utmost importance, by their very character they lack the factor of profit which is the essential motive of private enterprise. Where, then, shall they be undertaken? And how? Today only our large universities possess the necessary diversity of intellectual resources. In addition, they have accumulated in recent years invaluable experience in collaborative attacks upon the frontiers of progress.

All the examples that I have chosen have also this in common: their solution is imperative for the welfare of society. Investigation in any one of these fields will generate new insights and produce new knowledge and, to that extent, relate to the traditional mission of a university. Yet they extend far beyond the academic domain taken in its classical sense.

University centers for regional studies or for the economic development of new countries may be very scholarly in their character, but if their work is to be fruitful, it must link closely to the field of action. The ground, of course, is treacherous. Although a center for urban studies must not allow itself to become entangled in the politics of a particular city, the principles of urban planning cannot be developed in a scholarly vacuum. It is difficult to deal with the problems of the real world and remain aloof and untouched by that world. And so I think it inevitable that if universities engage in such undertakings, they must anticipate that step by step they will be drawn increasingly into a more direct participation in the active affairs of society.

In these remarks, one will recognize, of course, that I am dealing with just one facet of the broader development of the multiversity described by Clark Kerr. And although this development has become a reality for many institutions in the United States, the prospect of further change along these lines appears anything but enchanting to many of our colleagues in the academic world. There are many, both within and without the university today, who would like to reverse the stream and withdraw from the arena of action into the quieter, more scholarly domain that many of us knew in our younger days. Yet I submit that we are confronted by a most difficult dilemma. First, I must make clear that I am not considering here the relation of the university to local industry and its importance as a stimulating focus for the development of the regional economy. I have rather in mind areas of research bearing primarily upon the general well-being and the progress of our people. If the university declines to engage its forces in an effort to find solutions, how else shall we proceed?

Is it really possible to devise some wholly new kind of institution—a creation of our own age—that is neither an agency of government, nor a subsidiary of industry, nor wholly academic in its objectives? Many of the nonprofit corporations that have sprung up in recent years are exploratory efforts toward that end. Perhaps our generation will see some such organization take viable form. But it seems to me, in reflecting upon the problem, that it will prove very difficult for such an institution to survive over the long term and to become a permanent part of our society.

Consider for a moment the special qualities of the university. There is the tradition of scholarship, of intellectual excellence, that goes back for hundreds of years. There is a common ground for the exchange of ideas and a forum for scholarly discussion. There is the unparalleled diversity of interests, the arts, the humanities, the physical and social sciences, mathematics, and all the rest, which together constitute an academic community. In a more material sense, a university represents a certain stability. It has a history, a board of trustees or regents who understand its purposes and protect its interests. It has its own financial resources. And finally, and perhaps most importantly, it has an innate property of self-renewal—to use John Gardner's term—the self-renewal that comes through association with youth, the constant stimulation of students who for a period come under its influence and then move on to take their place in the ranks of citizens.

These seem to be indispensable elements in the constitution of an organization that is to survive and be fruitful. They are not easily reproduced.

It is a simple matter to establish new and independent centers of research. These may shine brightly for a time, but the test will come as they age. And only experience will tell whether they can develop those essential qualities that have given life to universities throughout all history.

I have been describing the paths by which universities have been led into new responsibilities and new commitments. The problems which increasingly occupy their attention are indeed central to the progress of our contemporary society. But they represent, too, one of the great intellectual challenges of our time. For the joint forces that are being applied toward solution constitute a powerful movement toward the synthesis of knowledge and a counter to fragmentation and the proliferation of a multitude of specialties. It is only within the framework of the modern university that the task of articulating or welding together the components of learning into systems of understanding can be successfully accomplished. And it is only through such an understanding that useful action will be achieved.

12

A New Order of Responsibility

The Commencement Address delivered at M.I.T. on June 12, 1964.

This is the first time in forty-two years that the President of M.I.T. has been called upon to deliver the Commencement address. I am immensely honored. Fortunately also I am untroubled by superstitions, for the last occasion of this kind on our campus ended in a terrible downfall.

By chance that was also on a twelfth of June. The Class of 1922 chose the Great Court for their Commencement exercises, and some two thousand people gathered under an enormous tent. That day was bright and clear, but during the morning the wind began to blow. And just as President Elihu Thomson, over the sound of flapping canvas, launched into his words of exhortation to the seniors, a tremendous gust struck the tent. The outer poles cracked, and slowly the roof began to settle down upon the heads of the audience. With flying robes and skirts, the faculty, the students, and their parents made their way out safely; and on the steps of the colonnade, under the dome, diplomas were finally delivered in delightful confusion.

Only seldom does an orator succeed in bringing down the house—and almost never at Commencement time. Today I hold out no such expectations. In this great structure, which for some bizarre reason is called the Cage, you may feel secure.

And now, in a more serious mood, I speak to you, who are

about to receive your degrees, not as a transient visitor on a strange campus, but as one who out of the experience of many years at M.I.T. has learned at firsthand what it means to be a student here. You have worked hard for this moment. I know the pressures that have been upon you; the seemingly infinite demands upon your capacity to learn; and the tremendous pace set by a quality of students that constantly challenges the faculty. I believe I know, too, something of the hopes of your families, of their pride in your achievement, and of sacrifices that have made your education possible. You have good reason to be proud this morning—all of you.

You came to us out of a most extraordinary diversity of origins and backgrounds. There is hardly a state in the Union which is not home to some of you, and you represent countries in every part of the globe. You have grown up in many faiths and traditions. You have known both wealth and poverty. You have attended small schools and large, public and private, in country districts and in great cities.

Yet for all these differences, you brought with you a certain basic community of interests and abilities. You were moved by much the same hopes and ambitions. Some of you were directed toward engineering or science; others toward architecture, toward economics or political science, toward psychology or management. But all of you, I suspect, shared the same concept of what M.I.T. stands for in the world today and viewed in the same light the role of science and technology in human affairs. It is upon that role, and how it is changing, that I should like to speak this morning.

Within the infinitely complex network of intellectual, emotional, and spiritual forces that make up this strange body we call mankind, one may discern two primitive impulses that trace back to the beginning of time.

The first is the desire to understand—the passionate quest for knowledge for its own sake—about the stars, about matter, about living things.

The second is the urge to do, the drive toward action and mastery of our physical environment.

The systematic interpretation of nature into a framework of law we call science; the effort to convert pragmatic experience and understanding to useful account is engineering.

For countless centuries the quest for knowledge through science has been moving forward, slowly gathering momentum, while the engineer has provided shelter, assured our supplies of food and water, built our roads and bridges, and created our massive industrial technology. And now, suddenly—almost within your own generation—the whole sweeping line of advance seems to have taken fire. In some strange unforeseen way, we have come to a critical threshold, beyond which the forces of technical progress appear to be self-sustaining. The processes of discovery, invention, and production feed upon each other. In every domain of the physical and biological sciences, there is a bursting out into new fields and new theories. The translation of ideas into action is taking place at an ever accelerating pace, so that the functional line of demarcation between scientist and engineer has almost vanished. From the factories and commercial laboratories of our country pours a mounting stream of new products, new versions of old devices—from jet airplanes to transistor radios, from nuclear reactors to household appliances, from a multitude of new drugs to synthetic building materials. We are at the point of being overwhelmed by the very bulk of our accumulated information, bewildered by the diversity of our manufactures. And we are failing today to assess clearly the implications of these developments for tomorrow.

Yet through this maelstrom of scientific and technological enterprise runs the almost mystical conviction that somehow every technical advance will contribute ultimately to the good society. Every responsible physicist believes intuitively or subconsciously that each new insight into the structure of matter will stir someone else—some engineer—to the development of a useful piece of hardware; and every engineer, in turn, expects that each new product or service will in some way add to our health, comfort, and material well-being.

There is nothing new in this idea of technology as the driving force of progress. It is an idea that took form during the Enlightenment of the eighteenth century and emerged from the Industrial Revolution of the nineteenth as a well-defined philosophy. Our own M.I.T. has its historical roots in that concept of the usefulness of science. Nor is there any need for me to demonstrate to this audience by examples how the advance of technology has improved the material condition of mankind.

Yet for all my own faith in and dedication to the work and methods of science, I do not believe that we can any longer afford to take such a thesis for granted; and I fear that a blind confidence in the inevitable good of material progress can lead only to disillusionment. The stupendous revolution of the twentieth century is doing more than adding theories, data, and apparatus to the accumulated store of the past. It has provided an entire new dimension to human affairs on a total change of scale.

To use words that you are particularly well qualified to understand, science, technology, and society now form a tightly coupled system. Each new technical advance adds a component to that system. In years gone by we have isolated these components and assessed their usefulness in terms of a specified purpose. We measured the value of a military weapon solely by a military requirement, a new drug by its immediate effectiveness in dealing with a particular pain, a new highway simply by the number of cars it carried, or a chemical waste disposal plant by the interests of local inhabitants. But now, because such components are coupled into an immensely complex system on a huge and massive scale, it is only by an examination of the impact upon society as a whole that we can pass judgment on the degree of progress.

To draw upon a biological analogy, I am saying that we must advance from the anatomy of components to the physiology of the organic whole—which, indeed, is now the society itself. One may prescribe an aspirin for a headache or build a turnpike to ease a traffic jam. But the headache may be merely an isolated symptom of a deeper disorder—a disease that may be identified only by a diagnosis which itself is the product of many specialists working together.

And so, too, our society—the body politic—is subject to old, chronic disorders and to new ailments. These diseases of the system are emerging in increasing number; and we must be courageous in recognizing that they are themselves the by-products of our highly technological environment.

Consider the transformation of our cities—the physical and social degradation of large areas—the loss of serenity and beauty. We have never before produced so many cars or such fast airplanes; yet transportation in the United States is rapidly approaching a point of crisis. The shift to automation in industry is accelerating and will have profound effects upon the character of our labor force, upon its training, and upon its security. We are polluting our air and our water. The pesticides which we are employing on a mounting scale are a boon to agriculture and a threat to the remainder of our natural resources. We find extreme poverty in the midst of affluence. The problem of the economically deprived citizen, be he black or white, is one of training and education to cope with a highly technological and rapidly changing society.

These are but a few examples. I could cite many more—all too large, too complicated, too sophisticated to be conquered or even arrested by sporadic investigations and isolated research projects. Of course science has generated these problems, and we can be equally confident that science can help us to alleviate and resolve them. But our efforts must now move to a higher plateau. We can no longer afford to nibble away piece by piece at the problems of the modern city, of transportation, of underdeveloped economies, of automation, or of disarmament. We indulge excessively in uncoordinated conferences, surveys, and studies that, on the whole,

117

are highly unproductive. Our ailments are vast and complex, and they will yield only to planned, collaborative attacks focused on clear objectives and leading to concerted action.

One of the charges that has been most commonly leveled against science is that progress is leading increasingly to the fragmentation of knowledge and the proliferation of a multitude of specialties. But these great new sociotechnical problems and the systems they represent are now also generating strongly countervailing forces toward new unities, bringing together many different resources, and giving rise to a new synthesis of knowledge. For in every instance, success will depend upon the joint contributions of physical and biological scientists, of economists and political scientists, of engineers and architects, of historians and philosophers. The task of articulating or welding together these components of learning into systems of understanding offers the highest intellectual challenge of our time.

As I have reflected on these matters—as I have pondered about how all this is to be accomplished and where—it has seemed to me that there exists nowhere at the present time any one institution or specific kind of organization which is in a position to undertake alone this monumental task. There is to my knowledge no single agency of government which has the necessary diversity of resources and the freedom of action. By their very character these problems lack the motive of profit which is the essence of private enterprise; and because they lead inescapably from the intellectual domain into the field of action, they present definite risks and perils for the university.

And yet it is only within the framework of the modern university that one finds the wide range of interests, a common ground for the exchange of ideas, a forum of discussion for scholars who draw upon the arts and the humanities as well as upon science and engineering. This is particularly true of an institution of the character of M.I.T. I do not believe that we can escape the responsibility of taking part in the solution of problems which touch most deeply upon the total welfare of our society. In the synthesis of knowledge—as well as in the creation of new learning —we must lead the way. And though at times we shall find ourselves drawn more deeply into the mainstream of contemporary affairs, we shall continue, in the process, to educate with relevance to our age. This has indeed been our historic mission.

I have wanted in these few remarks this morning to give you a broader and perhaps a new insight into the changing role of science and technology. The essence of this change lies not so much in the expanding scale of discovery and application; it lies rather in the complete penetration of science and technology into every domain of human affairs.

But principally I want you to understand that these developments have brought a new order of responsibility for the consequences of progress to our society as a whole. Some of this responsibility falls upon our institutions—upon government, upon industry, and upon the university. But in a free society the burden of obligation rests ultimately upon you as individuals. You are in a superlative position to meet the technical challenges of our day. But you owe something more to the common account. You must be ever mindful of your own deepening responsibility, both to the profession you have chosen to follow and to the society which will look to you for positive action. Your life will be productive in proportion to the goals that you achieve; rewarding in proportion to your commitment to all that is of value; and full in the satisfaction that the world will be better for your efforts.

13

Individual Freedom and Personal Commitment

The Commencement Address delivered at M.I.T. on June 11, 1965.

For many months I have had some thoughts in mind that I have wanted to convey to you this morning. But over the past week or two, as I have read newspaper accounts of one commencement address after another, there have been moments when I was sorely tempted to abandon the whole project. I have, in fact, felt very much like the final speaker at a banquet who listens with apprehension, wondering whether at the end, there will be anything left for him to say.

Since last September a series of incidents—one might almost call it an epidemic—on and off our university campuses has absorbed the interest of the American public. College presidents, in particular, have followed them with spellbound attention.

In both the North and South, students—sometimes as individuals and more often in organized groups—have taken an active part in public demonstrations on behalf of civil liberties. Student pickets have massed before the White House to protest a variety of causes. In one place or another they have demonstrated for free speech and against academic decisions on tenure and promotion. Not long ago they gathered in Pennsylvania to define the role of the faculty and the administration in the student-run university of the future. And then in that most novel innovation, the teach-in, students and faculty have joined forces to examine and discuss American foreign policy as it is revealed through our actions abroad.

All this—as you well know—has opened the floodgates of response and commentary. We have been treated to an impressive array of interpretations. Responsibility has been placed here or there, according to the sympathies or bias of the observers. But the force and depth of reaction have only become truly evident on the commencement platforms of the last few days.

One might almost conclude that everything possible has already been said. Yet I have decided to hold fast to my original subject, because it seems to me that it would be impossible in this June of 1965 to choose a topic more relevant to the interests of all of us who are here today. Moreover, these issues will not be disposed of quickly. The deeper we examine them, the more we come upon questions that are hard to answer. The views I express, of course, are personal, but inevitably the perspective with which I see the events of the day is influenced by the particular character and environment of our own institution.

We are, in fact, in the presence of movements, involving more than students, which reflect a change in mood on the part of many people throughout the country. However complex each particular incident may be, and however greatly it may be shaped by local circumstances, there run through each of them a few recurring themes. One relates to the right of personal freedom—hardly a new idea in itself.

The other evening, as I was casting about for an opening to these remarks, my Radcliffe daughter suggested that I have a look at a passage from Matthew Arnold. It is taken from an essay on *Culture and Anarchy* and is called "Doing as One Likes." Freedom, says Arnold, is "one of those things which we thus worship in itself, without enough regarding the ends for which freedom is to be desired. . . . Our prevalent notion is . . . that it is a most happy and important thing for a man merely to be able to do as he likes. On what he is to do when he is thus free to do as he likes, we do not lay so much stress." And then Arnold goes on to say, "this and that man, and this and that body of men, all over the country, are beginning to assert and put in practice an Englishman's right to do what he likes; his right to march where he likes, meet where he likes, enter where he likes, hoot as he likes, threaten as he likes, smash as he likes. . . . The moment it is plainly put before us that a man is asserting his personal liberty, we are half disarmed; because we are believers in freedom."

These words were written in England ninety-seven years ago. They suggest that perhaps times have not changed so much as we think.

Our own current preoccupation with freedom of speech, with individual rights, and personal liberties, extends, as I remarked earlier, far beyond the limits of the university campus. It is reflected in the views of those who assert that anything may be written in the name of literature and that anything may be presented on the stage or shown on the screen in the name of theater, and who insist that every restraint that might be imposed by good taste, if not by law, is an intolerable censorship.

We hear time and again that an individual has a right to do anything he wishes provided only, as they say, that "it doesn't hurt anyone else." I am wholly conscious of how many people today defend precisely such a position. But the question remains unresolved as to who shall judge whether an individual action—such as an indulgence in drugs, for example, may harm directly or subtly those around us.

Our country has been founded on the rock of personal liberties, and our Constitution was framed to insure them. Yet I cannot escape the conclusion that dangerous trends are appearing in the manner in which we interpret the intent as well as the content of our constitutional principles. Even our highest courts, in their concern for the technical rights of the accused, seem from time to time to forget that they are under obligation also to protect the rights of the victim and the safety of the public.

The very concept of *complete* individual freedom is in fact a fiction. It has never really existed. Primitive man in the forest lived without statutory law, but his freedom was threatened and limited on all sides by predatory beasts and human enemies. He gave up a measure of that freedom for the protection of an orga- nized society. The whole historical progress of civilization has come about through the step-by-step reconciliation of conflicting personal desires and the concession of certain individual free- doms. Civil and moral law are anything but a set of absolutes. They represent a succession of compromises for the sake of creat- ing an orderly world, wherein a degree of personal freedom must be yielded in the interests of a larger and more permanent free- dom for all. Codes of ethics, the statutes for the state or nation, the regulations of a university—all have developed out of this common need. They are more than often arbitrary; they are fre- quently in part illogical; they can rarely be defended on the grounds of any supreme absolute verities. But they are the frame- work which enables men and women to move in peace, to work in reasonable harmony. They represent the concessions we make not only to the body of civil law but to the common law of good taste and moral judgments. They are never perfect, and we should constantly strive to improve them and to adapt them to changing times. But a belief that each one of us individually is free to act at any time in accordance with his own personal in- terpretation of whether a rule or law or a canon of taste is right or wrong can lead only to one consequence: to an ultimately chaotic and disorganized society.

In these comments on the limits of personal freedom I have manifestly been speaking of views and trends that are by no means confined to the youth of our country. Let me turn now to a second idea, which comes very much closer, I believe, to the present college generation. I shall call it the need for a construc- tive purpose.

There are, as you know, wide disagreements about the basic causes that underlie the commotion and conflict of the past

123

months. To some observers they represent clearly an expression of idealism—and certainly there has been ample evidence of individual commitment, of the development of social awareness, of a desire to serve. But others see only restlessness, irresponsibility, the absence of concrete goals, the resort to action merely for the sake of action. And again there is abundant evidence of individual instances to support such a conclusion.

But wholly apart from such judgments, these varied movements, taken in their entirety, present a most serious issue to all Americans. We have been bitterly reminded this year of social inequities, of slums, and of poverty. And yet, on the other hand, no country in history has achieved our stage of technical progress, of material prosperity, of mounting wealth.

Many a historian has asked how long a society can survive its own affluence, can continue the expansion and deepening of its material life, and still maintain the sense of unity, of cohesion, of purpose which develops in periods of common striving. To me the hope for an affirmative answer to that question lies in our ability to capture the imagination of youth, to stimulate idealism, and to offer constructive goals. I am not wholly persuaded that this challenge can be successfully accomplished by diffuse projects and random actions alone.

The problem was clearly posed some sixty years ago by William James in an extraordinary address entitled "The Moral Equivalent of War." The burden of James's argument was that from time immemorial great wars have served the purpose of galvanizing youth into action, of creating moments of total national unity and direction. William James, you may remember, was an eloquent pacifist, but he acknowledged that wars do result in a national commitment of energies and resources toward a definite goal. He contended that in our efforts to do away with war for all time, we must find some equivalent to engage the forces of a nation in a great constructive cause.

Unhappily, no one has as yet come upon such a solution. Perhaps the nearest approximation on a modest scale is the Peace

Corps. The immediate and extraordinary response to that plan—and its success, contrary to the expectations of many of us—clearly demonstrates the genuine need and the validity of James's idea.

As I have continued to reflect upon these problems, I find it increasingly difficult to reconcile this new restless spirit of youth with the intent or aims of higher education. For Americans education has been the panacea—the ultimate cure for every ill. Generation after generation we have enlarged the base of public education, and we can be proud of what has been achieved. From an expected minimum of elementary school we have moved to high school; and now an attendance at college is taken more and more for granted. Is it not possible that we have concentrated so much on the idea of education as a good in itself that we have not thought enough about the ends to which education should lead? Enrollments are mounting year by year; but as one surveys the national scene broadly, one discerns that many undergraduates fail to develop a clear plan or educational purpose throughout their entire four years. I recognize, of course, the importance of college as a time to inquire into what the world of learning has to offer, and I recognize, too, that one should not be too precipitate in the choice of a career. Nonetheless—and very likely I am reflecting my M.I.T. bias—it does seem to me that there is the need for some focus, for a sense of moving toward some objective that should become ever more clearly defined.

Symptomatic of the present trend, I believe, is the increase in the numbers of college students who go on to graduate study without any serious commitment to some professional or authentic scholarly interest. It is just this deferral of commitment, this absence of an emerging purpose that leads to a sense of futility and frustration.

And so, finally, these thoughts on commitment and purpose bring me back to you who are graduating today. As I noted at the outset, your views and mine—students and faculty—are inevitably influenced in one way or another by the special character of M.I.T. This has been very apparent in the many discussions that have taken place about these matters on our campus during the past winter and spring. We have heard one opinion that our students tend as a whole to be insensitive to the political and social problems of the day. With this statement comes the call to a greater involvement in social action—notwithstanding the fact that M.I.T. students have contributed constructively in many ways to community projects. There is an opposite, and I believe predominant, view that the most important obligation of a student at M.I.T. is to take the fullest advantage possible of the resources available to him here and now and to direct his efforts chiefly toward the goal of a professional career. It seems to me that this is a choice which each individual must make for himself, and that he must be respected in his decision. In essence it is a decision on the fruitful use of time and a balancing of present and later values—a weighing of contributions that one may make now, at the expense of one's primary objectives, against later, perhaps more effective contributions made with the added power of professional competence.

The betterment of our society will come about only through the combined efforts of many kinds of people. In this common effort, by the very nature of the modern world, you know that science and technology are going to have a dominant part to play. There can be no doubt whatsoever of your ultimate obligation— as scientists and engineers, as architects and economists and managers—to accept your share of responsibility for the social and political life of your country. And in my own mind there is no doubt whatsoever that you who today become alumni of M.I.T. are indeed conscious of that responsibility and that by your achievements you will make us proud.

14

The Humanities in Professional Education

An address given on the occasion of the inauguration of Dr. H. G. Stever as President of the Carnegie Institute of Technology on October 21, 1965.

Some months ago, when I was invited to take part in the events surrounding this inauguration, the hope was expressed that I would comment tonight upon the relation of the humanities to professional education, particularly in the context of an institute of technology. I might have wished for an easier assignment, for the theme is hardly a new one. Since the founding of institutions such as ours fifty or a hundred years ago, the place of the humanities has been a lively subject of debate.

In many respects we were patterned after European models in the middle of the nineteenth century. But the idea that such subjects as English and history and modern languages should have a part in the professional education of engineers was distinctly an American innovation. It is significant that today, when nearly every country in Western Europe is engaged in a reassessment of higher education, they are examining closely the American experience and have shown a particular interest in the manner in which we have drawn the humanities into our programs in science and engineering.

The recent report of Lord Robbins, for example, proposes sweeping revisions in the plan of higher education in Britain.

Several years ago both Dr. DuBridge and I were invited to appear in London before the Robbins Committee, and I recall that the questions that were pressed upon me most strongly dealt precisely with this issue of the humanities.

The German technical universities also are currently in a state of ferment and change, and they have taken major steps in the integration of liberal studies into the scientific and technical curricula.

Without any doubt, the humanities are a real presence on the campuses of such institutions as Case, Caltech, Carnegie, and M.I.T. We make much of them in our discourses to the public. The concept is fundamental, and the intent has been sincere. Yet I think all of us must acknowledge that although a great deal has been accomplished and although the humanities have indeed had an impact upon our institutions, we still have far to go if they are to come finally into their own. The ideas that I present to you briefly this evening carry a certain sense of urgency. They express my own deep conviction that in the higher education of scientists and engineers, the humanities must be brought into full partnership.

I recall very vividly an incident that occurred several years ago at the Saturday Club, an old Boston institution with the traditions of Emerson, which describes itself as "a pleasant, utterly informal company of men, more or less eminent, having a long lunch together on the last Saturday of each month." The poet Robert Frost was there that day, and after a while the conversation turned to the teaching of humanities to scientists and engineers. Suddenly, to the dismay of almost everyone, Mr. Frost pounded the table and said to me with great bitterness: Let the scientists and technologists take care of their own affairs, which are going badly enough; let them play with their hardware, but let them leave the important problems of this world to us, to the poets and the philosophers.

Mr. Frost, of course, was not alone in his fear, and sometimes contempt, for the works of modern science. And it may very well be that he distrusted the scientists more than he did science itself. There are still those who believe in the desirability and even the feasibility of such a division of the world's labor. Yet this idea does injury to the cause of the humanist as well as to science and runs counter to reality. For whether we will or not, science cannot be isolated from the rest of human affairs. Nor can anyone understand the stream of art and literature and philosophy today without some perception, some awareness of the great currents of scientific thought. We are a single society, a single world, with a single complex of human problems. And science, in effect, is simply the search for comprehension, the attempt to unravel some part of the mysteries of the universe which we inhabit.

Contrary to a very common belief that the progress of science is leading us toward a fragmentation of knowledge and increasing specialization, I see clear evidence of a growing confluence of fields and disciplines. Today, for example, it is extremely difficult to draw a line of demarcation between what constitutes pure science and modern engineering. It is difficult to isolate physics from chemistry, or chemistry from biology. And engineering is itself a continuum, extending into economics, political science, and management. But this flowing over into broader areas goes much further. By the very nature of their wider concerns and the character of life as it is today, scientists and engineers are drawn deeper and deeper into issues and decisions whose import is far more than technical. And it is for such responsibilities, for this new order of citizenship, that we must prepare the graduates of our institutes of technology.

At the heart of this matter is the ideal of the true professional estate. Let us think for a moment not about scientists or engineers, or about lawyers or doctors, but about the qualities that should distinguish the professional man or woman.

First, of course, there is the essential requirement of complete competence, of a depth of understanding about some special field at a high intellectual level. Just the other day I came across as clear a statement of this requirement as one might wish. It appeared over ninety years ago in the annual report of the Head of the Department of English at M.I.T.

"The professional student," he said, "comes for a distinct purpose; he wishes, for instance, to be made an engineer, and he must be trained so thoroughly in engineering studies that his bridge will not break down through faults of construction. It would be small comfort to his employers if that should happen, that he could report in Addisonian English, and with unimpeachable logic, the precise reasons why it broke down."

This kind of mastery implies more than a command of current practice. It calls for such an understanding of fundamentals as to resist obsolescence and to instill a confidence in one's own power to maintain a position in the forefront. But beyond the requirement of commanding knowledge and adaptability to change and innovation, the professional estate is a way of life. It reflects a point of view—a concern for consequences beyond technical terms; a comprehension of the impact of one's works on the intellectual and spiritual life of the individual and of the society; a willingness to administer to the public welfare; a sense of civic responsibility; an understanding of our institutions that derives from a sense of history; a taste for excellence and a capacity to recognize and to appreciate the beauties of ideas, of art, and of literature. It calls for a commitment to the ethical principles on which decisions must rest. It is expressed through a reaching down for fundamentals with a desire to get to the essence of the matter, and, above all, to an understanding of one's own relation to what has gone before and what is to come—in sum, a human perspective. Admittedly, these are virtues that few men ever achieve completely, but ones to which we hope our graduates will aspire.

It would be unreasonable and ill-advised to assume that the faculties of English, of history, of philosophy, or of languages in our technical institutions can alone develop and instill these qualities in our students, or that the burden should be solely theirs. Yet from time immemorial, we have looked to the humanities for inspiration and leadership in just this realm of ideas. And if institutions such as Carnegie and M.I.T. are to fulfill their responsibility for professional education in this highest sense, the humanities must be more than peripheral. They must become central to our endeavor. It is in this perspective of a truly professional education that we must give them encouragement and support and that they, in turn, must respond to the challenge.

And now how are we going to bring this about? It will not be easy to achieve, nor is there in my judgment any unique way to augment the strength and consolidate the position of the humanities in our professional schools. We shall have to try many experiments and explore many roads.

While all these institutions have much in common, they also have different histories, different resources, and different manners of approach. Carnegie is fortunate, for example, in its College of Fine Arts, and it may capitalize on the extraordinary opportunities for the enjoyment of art and music offered by this city of Pittsburgh.

Of one thing I have become quite convinced: it will be difficult, if not pointless, to attempt to recreate a miniature of a traditional liberal arts college as an enclave in the midst of an institute of technology. The humanities must not be set apart. If they are to thrive, they must put down deep roots of their own for an authentic growth in our peculiar environment. What we seek is a style for the institution as a whole rather than the cultivation of a secluded little plot of ground.

This, in turn, implies new forms and demands new insights and imaginative solutions to new and old problems. I have no thought this evening of discussing details of administration or specific programs—the pros and cons of core curricula or course sequences—but I do want to emphasize the need for a revolutionary attack. My colleague Elting Morison has remarked that we must recognize that "unlike most universities, we are less custodians of what has been said and thought and more creators of the new. What goes on in science and engineering is a restless seeking for the undiscovered or untried. This spirit should be made to invest our work in the humanities."

At the same time, I must warn against what someone has called the "scientization" of the humanities. Linguistics, mathematical logic, and the quantitative applications of psychology are important in their own right and, by their very nature, are strongly linked to mathematics, to physics, to the communication sciences. The history and philosophy of science are also eminently germane to our interests. But these are not themselves humanities in the sense of which I speak, nor do they fulfill our need. We may fruitfully draw upon the modern world for examples and stress the society of our day rather than of antiquity. But the essence of humanistic studies is rooted in the timeless issues of mankind.

We must always face the question whether realistically these institutes of technology can vie successfully with the colleges and universities of the older liberal tradition in building strong faculties. I think they can, and I am confident that they will.

Of course the arts here will not be the same. They will develop their own form and character. It would be absurd to imagine that all areas of literature, for example, would find scholarly representation among us any more than one might expect to find every specialty of science in a liberal arts college. And yet I think we have much to offer to the scholar and the artist who will come to terms with science. There is a vitality in our institutions. There is a sense of involvement, of relevance to the world around us, to the problems of our time. There is a striking sense of forward movement. And there are students of superb intelligence, eager, extraordinarily receptive, open to ideas. This, I submit, is anything but sterile ground for humanistic studies.

Science and engineering today are the very matrix within which human affairs must be interpreted. As we speculate about the destiny of mankind, we draw upon what science teaches us of the origins of our earth and of the biological nature of life. As we strive to develop that destiny, we draw upon the power and implications of modern technology. As we seek to interpret that destiny, to move mankind forward in understanding and perception, we look with hope to the working partnership of the humanities and science in its broadest sense. As the greatness of the Renaissance lay in the meeting of the scientist and the humanist on common ground, so we must aspire to create together the man of the twenty-first century.

PART II

M.I.T.: INSTITUTIONAL AND PERSONAL EXPERIENCES

15

A Technological University

A lecture broadcast by The Voice of America in its Forum Lectures series during 1964.

I have been invited to tell you in this lecture something about the Massachusetts Institute of Technology and how it has come to assume in recent years the character of a modern university. I shall speak of it simply as M.I.T., for that is the name by which it is known around the world.

M.I.T. makes no formal claim to the title of a university; yet no other word designates more accurately the totality of interests and objectives it encompasses today. What, then, are we to understand by this term *university?*

There exists, of course, not just one idea about what a university is or should be, but many; and the modern university is indeed many things. All universities, however, have certain general purposes in common. They are the means through which we preserve and interpret the accumulated learning of the past and transmit this heritage to each successive generation of students. Through the arts and sciences they undertake to strengthen the liberal culture of the individual. Universities are distinguished from colleges in that they provide not only the foundations but also the specialized training for a number of the higher professions. And every true university is committed to research—to the process of adding to our total store of knowledge and thus contributing to the advancement of learning.

But no single institution today, however old or however large, can seriously boast that it takes all learning for its province. The accumulated wealth of modern knowledge is too immense. And so each university must to some degree mark out its own domain, focus its efforts, and develop its strengths in areas and within limits prescribed by its own circumstances.

In one further respect the contemporary university—particularly in the United States—departs radically from the classical idea of the past. Few institutions of higher learning still preserve the character of the ivory tower, the lofty isolation from the great political and social movements of the world about them. On the contrary, they are increasingly drawn into the life of the times and are developing countless ties with their local communities, with industry, and with their national governments.

Fundamentally, the university is a social institution. Wherever it may be, it draws upon the history and traditions of its own land and people. Its form, its outlook, its procedures in each country are a unique expression of national character, and it must be built to serve national needs and aspirations. The truly essential concerns of a university are for new ideas, for new knowledge, and for youthful minds. In the world of today the university has consequently become a major instrument of social progress, and it must equally be alert and responsive to change in the society of which it is a part.

Now of all the forces that are shaping modern society certainly the most powerful spring from the tremendously rapid advances of science and engineering. As a result, the education of scientists and engineers has become, in every country, an objective of the highest priority. It is this very fact that gives to such institutions as M.I.T. today their special meaning and importance.

For an understanding of how this meaning has developed and of how the character of M.I.T. has changed to keep pace, I must take you back for a moment over a period of about one hundred years.

In the middle of the past century, the United States was still an underdeveloped country. In the eastern regions large cities were beginning to rise, and the economy was turning from agriculture to industry. But the vast plains of the Middle West and the far lands of the Pacific coast remained virtually untouched. There were forests to clear, fields to plant, roads and bridges and schools to build.

In the earliest years of our republic we drew heavily upon Britain and Western Europe for the basic ideas that underlay our laws and institutions. Later, as a distinctive American character took form, we departed freely from our inherited or borrowed models and began to develop clearly indigenous institutions.

In 1850, or thereabout, the principal universities and colleges of the United States reflected in their curricula and in their organization the profound influence of their prototypes in Great Britain—chiefly Oxford and Cambridge. But two great movements of revolt were beginning to stir.

The first of these came from Britain itself. The utilitarians—Bentham, John Stuart Mill, Robert Owen, and their circle of friends—had been advocating the case for progress and science. Their faith in science, indeed, surpassed our own; for they were confident that in only a matter of time science would reveal to us all the mysteries of the universe, would satisfy the material wants of mankind, and bring peace and harmony to the world. Out of this emerged a powerful belief in the importance of useful knowledge. Soon there followed an attack upon the whole plan of British higher education. By mid-century we find Cardinal Newman defending the traditional idea of a university, maintaining that its purpose was instruction rather than research, its function to discipline the mind rather than to diffuse useful information. Matthew Arnold eloquently defended the humane letters as the single path to culture. But against the ancient fortress of classical learning, Herbert Spencer, Thomas Huxley, and many another pressed the cause of science and spoke for an education relevant to the problems of the day.

In the United States these radical ideas found fertile soil. The Industrial Revolution had come to North America. A system of colleges founded on rhetoric, Protestant theology, and the culture of ancient Greece and Rome, with the objective of preparing ministers, lawyers, gentlemen, and scholars, fell short of the needs of the country.

Innovation does not come easily—not even in the academic world. But out of the ferment and acrid debates of the nineteenth century came two new forms of educational institutions on this continent, authentically American in character and expressing a faith in the dignity of useful studies and a belief that they may be interwoven with a liberal culture. The first innovation was the great system of state universities and land-grant colleges; the second, the several institutes of technology, of which M.I.T. is an example.

M.I.T. was founded in 1861 by William Barton Rogers, who earlier had been a professor of geology at the University of Virginia. In his plan for the Institute, Rogers was influenced by the utilitarian ideas of his time and moved by the urgent practical needs of his country. He also drew freely upon British experiments with new schools and was thoroughly familiar with the programs of the Technischen Hochschulen in Germany and of l'École Polytechnique in France.

According to its charter, the Institute was established for the advancement and development of science and its application to industry, the arts, and commerce. The early curriculum was organized into several departments of engineering and a school of architecture. From the very beginning it was dominated by two basic ideas:

First, that the education of the engineer must be thoroughly founded upon a mastery of physics, chemistry, and mathematics.

And second, that the education of a truly professional man or woman must embrace liberal as well as technical studies. It is noteworthy that M.I.T. at the very outset offered subjects in literature, philosophy, and the modern languages, and even for a time one or more complete curricula that combined literature or philosophy with general science. Later this emphasis upon the humanities diminished, only to be resumed strongly, as we shall see, in recent years.

For more than sixty years the basic plan of engineering education remained virtually unchanged. One generation of students after another graduated from the Institute and went out into the world of affairs to contribute its share to the development of its country. The civil engineers built roads and railways and cities. The mining engineers developed the mineral resources of our West. The mechanical, chemical, and electrical engineers designed the tools, the machines, the processes of manufacturing, the fabulous complex of our expanding industrial power.

Education at M.I.T., from 1870 to 1920, was immensely effective, and the secret of its success lay in its relevance to the pressing needs of the country. But no institution can afford to stand still. By 1920 our curriculum had in its turn become rigid, a frozen stereotype of past practice. The same criticism could be made of almost every engineering curriculum in the United States at that time. The relative emphasis on practical work—mechanical drawing and shop practice—had increased. Physics, chemistry, and mathematics were presented not for their own sake but as "service courses"—with examples strongly slanted toward practical use. And studies in the humanities were reduced and diluted. The product of that system was a graduate who was immediately useful for industrial operations but ill prepared to push forward the frontiers of his profession.

There are tides in the history of institutions as there are in the lives of men. The decade of the 1920's marked the end of an era. The Institute enjoyed a well-earned reputation for its high standards, for the rigor of its methods, and for a single-minded concentration of effort upon the education of engineers. But the old pattern—the plan that had been enormously effective for so many years—was wearing thin and had lost its lustre.

I have commented earlier in this broadcast on how the development of a university reflects also the development of the society of which it is a part.

In the United States the outstanding intellectual movement thus far in the twentieth century began after the First World War with the cultivation of pure science on a broad scale. Before that time this country could offer only a few isolated examples of distinguished work in mathematics, physics, chemistry, or biology. But beginning in the twenties, American graduate students went abroad in large numbers to undertake research in such great centers of learning as Göttingen, Berlin, Copenhagen, Zurich, and Paris. They came home to establish fundamental science on a firm footing in our American universities. How science has prospered here in the intervening years is aside from my subject, yet I think it clear enough, whether one takes for evidence the number of Nobel Prize winners or the vast quantity of scientific papers that issue annually from our academic and industrial institutions.

It was the impact of this sweeping movement toward basic scientific studies that, in the 1930's, began the first great transformation in the character of M.I.T.

Briefly, the argument for change was this. The time lag between scientific discovery and useful application is rapidly growing shorter. Science and engineering are becoming completely interwoven. They form a continuous spectrum of effort rather than two distinct fields of endeavor. The technology of today may be obsolete tomorrow, and no one can foresee what discovery of physics or chemistry may generate a whole new domain of industrial production. Consequently, if the engineer is to meet the future demands of a highly sophisticated and increasingly complex industrial world, if he is to lead the advance and give pace to innovation, then his command of at least the fundamentals of modern science must be complete.

This position differs from the original plan of the Institute not in principle but rather in degree. The conviction grew that a proper climate for the study of science could be achieved only if pure science were cultivated for its own sake, with all that this implies in the context of modern scholarship. And so, beginning about 1931, we saw the systematic strengthening of the departments of mathematics, physics, chemistry, biology, and geology. These in turn were formally incorporated into a School of Science, the counterpart of the earlier Schools of Architecture and Engineering. Two other steps were taken at the same time to fortify these developments. A Graduate School was formally established, and graduate study leading to the master's or doctor's degree was rapidly expanded. Second, there began the systematic encouragement and support of basic and applied research in all fields within our province.

A new vigor inspired by new ideas and new aspirations was stirring M.I.T. at the advent of the Second World War. To my audience in countries throughout the world I need not try to express what that conflict meant for academic institutions in every land. Ultimately, with the return in the late 1940's to more normal conditions, there emerged at the Institute a wholly new perspective upon the role of science and engineering in the modern world and of our responsibilities toward education.

Throughout the earlier years of our institutional history, we fulfilled our purpose and our obligations simply by the graduation of engineers, architects, and, later, of scientists. Their task in turn was to build, to design, to investigate. How the product of their efforts might affect people, or what the impact of their works might be upon the society, was commonly thought to be someone else's concern; and the social implications of science and engineering were hardly considered to be a major concern for an institute of technology.

All such ideas of professional self-sufficiency, of remoteness from social consequence, were swept away by the events of the war and by the extraordinary progress of science that followed. Science, in the broadest sense, now exerts a penetrating influence upon art and industry, upon every aspect of our economic, political, and social life. The overriding practical problems of our time —disarmament, the economics of change, the politics of peace, the relationships among industry, science, and government—call for the joint application of technical and social analyses.

M.I.T. has accepted what it believes to be a new order of responsibility and has expanded its domain of activity to include the social sciences, the field of management, and a greatly increased emphasis upon the humanities. The result of these efforts has been to impart to the Institute a wholly new academic dimension.

The course of this development has followed lines and ideas similar to those that governed the earlier strengthening of the physical sciences. If, for example, we are to offer the student of engineering excellent opportunities for acquiring a sound background in economics, then the department of economics must fulfill more than a service function. It must be represented by a faculty strong in its own right and pursuing the subject of economics for its own sake. This principle, in turn, carries with it a wholehearted commitment to graduate study and to research.

To give substance to these intentions a School of Humanities and Social Science was formally established shortly after the war and now includes the faculties of three departments. The Department of Economics and Social Science comprises economics, political science, psychology, industrial relations, and sociology. The Department of Humanities serves those members of the faculty concerned with history, literature, music, and philosophy. The Department of Modern Languages provides, in addition to instruction in such languages as French, German, and Russian, courses in foreign literature, comparative literature, and linguistics.

In most of these areas the resources for study and research are extensive. Thus, the faculty in Political Science offers a very substantial number of subjects, all of which are related to other interests of M.I.T., in six broad fields: international relations and foreign policy, political communication, defense policy, government and science, political and economic development, and comparative politics. The Political Science group is closely allied also with M.I.T.'s Center for International Studies.

The subject of political and economic development has been one of particular interest both to this Center and to the Department of Economics and Social Science. A substantial number of graduate programs and much significant research bear directly on the development of the world community and on the establishment of peace. Individually, members of the faculty have served the United States government in important roles, particularly as advisers and consultants for the Agency for International Development and for the Alliance for Progress. They have also contributed to various advisory groups of the United Nations and of several countries, where they have helped to formulate and put into practice major development plans. Their scholarly interests cover a wide span, ranging from fundamental studies of the problem of arms control to the complex relationships between economic growth and technological change. And in all this work they have sought to bring a closer alliance between the natural sciences and engineering and broad social, political, and economic concerns.

The work of the Psychology faculty is focused on three large and important areas: physiological and comparative psychology, experimental psychology, and social and developmental psychology. These, too, are linked to other fields, from biophysics to industrial management, at the Institute.

Our entry into the field of linguistics was stimulated initially by the pioneering efforts of Professors Norbert Wiener and Claude Shannon of M.I.T. on the mathematical theory of communication. The study of the logical relationships within languages employs mathematical techniques derived from the general area of communication theory. Since our emphasis has been on the structure and logic of language and on the fundamental nature of oral communication, it is not surprising that our linguists have worked in close association with mathematicians, electrical engineers, and physicists as well as with biologists and psychologists—a superb example of interdisciplinary collaboration in actual practice.

Still another example of this same interdisciplinary trend in contemporary scholarship as well as of the evolving character of the Institute is the spontaneous growth of a strong group in philosophy. The current interest of philosophers in the older exact sciences—especially mathematics and physics—and in such newer fields as psychology and linguistics, has attracted to M.I.T. a most distinguished faculty concerned with the interrelations between philosophy and science. Their presence in turn has drawn to the campus leading scholars from the more traditional areas. Our work in philosophy, as it is now emerging, will concentrate on logic, epistemology, philosophy of language, and moral philosophy. And these studies will be linked most closely with others at the Institute.

In the present environment, the task of developing strong programs in the social sciences and in such specialized fields as the history of science and technology is relatively easy, and they are in fact flourishing. The problem of cultivating the humanities is admittedly more difficult, but we believe it to be urgently necessary for our purpose. All undergraduates under the current plan follow a basic curriculum of humanistic studies throughout the first two years. The time allotted is the same as that for mathematics. A comparable requirement is imposed for the upper two years, but the student may begin to concentrate and to pursue his own interests by a choice of three subjects from any one of ten fields including history, philosophy, literature, modern languages, music, the visual arts, or a social science. Furthermore, every encouragement is given to the undergraduate to add elective subjects of his own choosing.

The number of undergraduate majors in the social sciences is steadily increasing. Moreover, there has been a most gratifying increase in the election of subjects in the humanities and social sciences on the part of students majoring in science and engineering. But the most significant innovation in recent years has been the development of a four-year, highly flexible, undergraduate major which combines in about equal measure the humanities with science or engineering. The course has attracted excellent students who seek a broad but rigorous preparation for further professional work in science, engineering, the humanities, law, or medicine.

These are, for us, pioneering efforts. We have been guided throughout by the principle that the highest qualities of excellence must prevail in whatever new field is undertaken. The student of physics or mechanical engineering responds immediately to a course in history, or literature, or political science that commands his respect and imposes upon him the same standards of achievement to which he has become accustomed in his own professional field.

I have, in this account, had time only to touch briefly on a few new developments by way of illustration. If you were to examine the current M.I.T. catalogue, you would find there an impressive array of subjects offered in areas seemingly far removed from the original interests of an institute of technology. But the truly significant change is more subtle than anything which emerges from the pages of a bulletin. It is a change of spirit, of intellectual climate that pervades the entire institution. Nothing that has happened at M.I.T. over the past decade has in any way diluted the rigor and high standards of motivation and research in science and engineering. But these are now viewed by the faculty in the context of a larger mission. The humanities, the social sciences, the field of management, have assumed a central place in our affairs. The members of the faculty concerned with these studies no longer serve simply in the role of a supporting cast; they have become full and equal partners in the academic life of the Institute. The real force of the new movement lies in the growing unity of purpose and understanding.

I began this lecture with some comments upon the definition of a university. It is highly unlikely that the Massachusetts Institute of Technology will ever abandon a name that has earned such respect through the years. Yet I hope that you will associate with this name the interests, the liberal outlook, and the character of a modern university.

The objectives of M.I.T. have been more modest than those of some of the great universities of the world. We have taken science to be our central province. We have been preoccupied with the advancement of science and, through engineering, with the uses of science to enhance the welfare of mankind. But now we acknowledge also a responsibility for the larger influence of science and engineering upon society—for their impact upon the national economy, upon government, upon international affairs. And therewith we no longer find it possible to define and delineate sharply the boundaries of our concern. We have moved into an intellectual domain that is immense, embracing or touching ultimately upon almost every conceivable human activity. Yet the horizons of opportunity are expanding too rapidly for any one institution to stretch to them all. The most perplexing problem confronting M.I.T. today is how to select among new fields, how to respond to opportunity, and yet how to retain, within this broadened scope, the unity of thought and action that has given it strength in the past.

We live at a time when some look to science as the one great hope of the future, and others see in it only a force destructive of spiritual values. M.I.T. is an institution built upon science and engineering. Here, if anywhere, one should look for commitment to the concepts of science and faith in the power of its methods. But it would be a serious misreading of the temper and purpose of M.I.T. to conclude that our vision of life is no larger than the satisfaction of material wants, that we have no hope of peace other than through an arsenal of weapons, or that we are guided in our own lives by no principles other than the physical laws of nature. The whole evolution of the Institute belies that narrow view; the actions and interests of our faculty, their intellectual and spiritual concerns, speak eloquently for themselves.

16

The Realm of the Spirit

An address given at the M.I.T. Chapel on February 16, 1966, as part of a program arranged by the religious counselors of the different faiths. The program was held in Dr. Stratton's honor, and a presentation was made to him by the students.

A little more than ten years ago we built and dedicated this chapel. For mankind in every corner of the earth, this has been a decade of extraordinary development and change, a time of powerful movements. As one looks back over these years, it is very easy to single out the great currents that are sweeping us forward into a new era.

First there has been the massive progress of science. No one could be more fully aware than you here this evening of the tremendous advances that have been reported in every field. Step by step we seem to be closing in upon the great mysteries that have perplexed the minds of men for generations. One great pathway of research is leading us nearer to an ultimate explanation of the strange particles of physics, and so to the constitution of matter. Another, through biology and the new genetics, may take us a little closer to an understanding of the nature of life. And a third—a very exciting road into outer space—is opening to our view the wonders of the cosmos, of other worlds and planets, and of new laws of the universe beyond all our previous comprehension.

Then, in a wholly different domain, we have witnessed in our own country these past few years a sudden, a most extraordinary awakening of conscience and concern for civil rights. This has been a movement of almost explosive character, an outburst of indignation and action expressed in a national resolve to attack injustice and to make good on promises now more than one hundred years old.

Within a decade, we have seen abroad the rising aspirations of new countries, their achievement of independence, their struggles with the problems of self-government, and their continuing, intense efforts toward a better life.

And again here at home, in this land of enormous material affluence, we find ourselves engaged ever more deeply in conflicts abroad, striving to reconcile within our own conscience the desire to maintain and defend our liberties and the yearning for a lasting peace among the peoples of the world.

Indeed, whatever pride and satisfaction we may derive from our material and technological success must be tempered by a recognition of so many shortcomings—our failure to achieve any adequate measure of the equity, the justice, the peace, and the serenity that give dignity and meaning to the human drama.

And this brings me finally to still another movement that has been gathering force over the past decade, a powerful new current that in its historical context may yet prove the most significant of all. It may well be that our very shortcomings and failures —and the sharpness of their shadow against the background of our material progress—have given new directions and infused a new vitality into the role of the church in its widest meaning.

We see it through tangible efforts to meet the challenges of a constantly changing present while keeping faith with the traditions of the past. We see it, too, in a deeper and more direct involvement on the part of the clergy in the most urgent moral and social problems of our day, in a looking outward toward the community, in a freer participation with one another in the solution of common problems. From this willingness and understanding, from this growing involvement, the old phrase "the church militant" takes on a new significance.

At the chapel dedication ceremonies ten years ago, Dr. Ferris of Trinity Church spoke of a revival of religion that might be more than the rediscovery of old truths forgotten or neglected. There is every sign that such a revival is fully under way. With it, I think, has come another, perhaps unforeseen, development. For it seems to me that this concern for the universe in contemporary and in future terms—transcending a preoccupation with the past—has begun to bring about a working partnership among the representatives of many faiths. In turn, this has contributed in a subtle, yet enormously important, way to the whole ecumenical movement. It was the vision and deep human understanding of Pope John that brought this movement into being, but the extraordinary response throughout the world reflects a certain degree of preparation and spiritual readiness.

I like to think that in its own special way the concept of this chapel, where we meet tonight, foreshadowed this extension of common understanding. Even a short ten years ago the idea of one chapel for all was novel, even objectionable to some. Today we have come to accept it as wholly understandable, even desirable. It has in every way fulfilled our expectations. Not only for the novelty of its design are we proud to show to visitors the chapel on our campus; we do so also because it stands as a symbol of our concern for all those ideas that lie beyond the purely material aspects of life.

Several years ago at a Christmas Convocation I undertook to affirm the place of these ideas at M.I.T. My words at that time expressed so clearly my own convictions and faith that I am going to recall them briefly here again.

Our generation was brought up in an age that, more than any other, has been dominated by the ideas of science. We here are part of a community that has contributed and continues to contribute enormously to the advancement of science itself. And so inevitably our way of life, our manner of thinking, the values we attach to ideas are profoundly influenced by scientific principles and methods. Out of this environment there grows an attitude of mind that tends to place the highest values on that which is new and freshly discovered. The planets still move in their orbits according to the laws of Newton, but that simple concept has long since lost much of its excitement, though none of its beauty. Our minds are captured by the unresolved problems of physics, of biology, of astronomy. This, after all, is the very essence of research.

Daily we live and work in the realm of the intellect. Our approach to every problem is based upon reason. We are concerned with the rational processes of the mind. We deal in large measure with facts and figures, with elements that are tangible and concrete. To these traditional concerns at M.I.T. we have added, increasingly in the past decade, a leavening of the arts and the humanities so that we may become truly professional and abundantly aware of our heritage of ideas, of the sweep of history, of the beauties of art and literature. For these give balance to the range of our knowledge and understanding and a broader import to the works of science and technology that will constitute our own special contribution to the progress of man.

But in our particular concern for the things of the mind, it is well for us to remember that there is in addition the realm of the spirit. And it is precisely from this side of our being that have come the beliefs and the universal truths that sustain and uplift us.

The existence of this chapel at M.I.T. and our support of the program of the religious counselors underscores our commitment. If, on the one hand, the symbolism of the structure may seem to imply a withdrawal into an inner life, an escape from the real world, if we must cross a bridge to enter—to pray alone or with others, or merely to reflect—then we can be sure that this same bridge will nonetheless lead us back to the society of which we are a part.

I acknowledge the obligation of the church to be in and of the world, but it seems to me that it must be more than an agency for social action. We must never lose sight of its most singular quality, of its essential meaning. Each of us will define this meaning in his own way and according to his own belief. But we shall have missed the mark if we fail to preserve the wonder and the mysteries that transcend its outer form. This campus of ours is anything but a wasteland in which the spiritual may wither and die. It is, on the contrary, a most fertile ground for the new spirit of questioning, for probing further into the meaning of science, its expectations and limitations, and into the ultimate meaning and purpose of life.

Herein lies the greatest responsibility and opportunity for science and technology in the decade to come. We recognize the need. We acknowledge the duality of our lives. And we have taken major steps forward.

17

The Idea of a University

The address to the students at M.I.T.'s Freshman Convocation on September 15, 1965.

You are meeting here this afternoon for the first time as the M.I.T. Class of 1969. For each of you this is a supremely important occasion, one to which you have been looking forward for many months. It marks the beginning of a wholly new experience—an experience that you and we, working together, are going to make constructive and rewarding.

These next few days will be filled with all the activities of Freshman Weekend: registration, moving into your dormitories and fraternities, meeting new friends, talking with faculty advisers, finding your way about the campus. But then next week you will begin to settle down. You will attend your first classes. You will discover the tempo and scale of our life here. And, most important of all, you will begin to be a part of M.I.T.

Let me say at the outset that your place here will not be that of a freshman isolated with other undergraduates in a distinct, separate college. Rather, you will join with the graduate students, the postdoctoral scholars, and the faculty—with people from every part of the country and from all over the world—to make up this single, tremendously interesting academic community and to live and work in a shared environment of scholarship and learning.

Other generations of freshmen have come here before you. For a century, in the particular context of their own time, these other freshmen have encountered problems and experienced feelings quite similar to yours. They have met and conquered—just as you will—the pressures, the passing moments of uncertainty and discouragement. They have seen the four years—which appeared endless at the outset—slip by with incredible speed. And they have known—just as you will—the immense satisfaction and pride of success and achievement at the end.

You have studied very hard for the privilege of entering this institution, and I am certain that you have already given much thought both to your own goals and to your expectations of us. But today I ask that you take a broader view, that you look beyond the specific profession you hope to follow, whether it be some field of engineering or science, architecture, economics, political science, or management. Look beyond the facts that you hope to add to your store of knowledge and the schedule of classes that you must faithfully pursue. Consider in this new setting the deeper meaning and the implications of the years just ahead.

First of all, as you make the transition to M.I.T., you must begin to accept a new order of personal responsibility. To my mind, in this higher degree of responsibility lies the fundamental difference between secondary school and the idea of a university —a difference that is frequently misinterpreted, if perceived at all. At the moment you may think of the next four years as a prolongation—a kind of stretching out—of your high-school days. Yet, rightly understood, they should be very much more. For the nature of the educational process is not at all the same. True, you will recognize some of the familiar routine. There are required courses; there will be term papers, deadlines, exams, and grades. But a university offers you resources for learning beyond anything that you have known before. And the responsibility for the use that you make of these resources, the effectiveness with which you draw upon their multitude, shifts today from the teacher to you. You enter now into a mature domain in which the choices basic to your success or failure will be largely your own. Henceforth, you must assume a new measure of accountability for your own efforts and achievement and for the decisions that you must make on your own behalf.

This is not to say that you set out on this venture alone. I can assure you with the utmost sincerity that you will find a faculty that truly cares about your welfare and your progress and that is eager to guide you on your way. As I look back over the long years of my own experience at M.I.T. as a student and teacher, it seems to me that there has never been a time marked by a deeper or more active concern for the undergraduate, his problems, and the substance of his education.

You will find also that the vast resources of M.I.T. are those of a dynamic institution responding to the demands and the issues of our age. You need only take the general catalogue in your hands to have proof of the extraordinary range of opportunities that awaits you. There is freedom, if you exercise it, to shape and modify your course as you proceed. Many of you may, in fact, finally settle upon a career quite different from the one you have in mind today. This very real flexibility of which I speak is evident in the core curriculum, in the freshman seminars, and in the electives. You will find it in the chance to work in laboratories and, later, to participate in research. I urge you, once you stand on solid ground, to avail yourselves of this almost infinite array of possibilities. These are indeed the years to explore new fields of learning; and your acceptance for this entering class is in effect an invitation to do so.

But it is not a mandate to roam at random and without focus. That would be contrary to the spirit of M.I.T. My charge to you is to seek with an open mind but in a purposeful way, to reaffirm an interest or perhaps to discover a new one, to make it your own, and gradually to form your commitment. In doing so, you will develop your own style, and you will begin to design a way of life. You will work with us thoroughly and in depth, laying the foundation of a profession and creating a solid core of competence on which to build through all the years to come.

All the years to come—in this phrase is the key to the real significance of your stay with us. The very fact that you have sought admission to this particular institution makes it almost unnecessary for me to point out the fundamental role of science and engineering in our time. Yet I want to emphasize that, as their influence has deepened, pervading every aspect of human affairs, it has brought new meanings and new dimensions not only to life but, in turn, to the whole of education.

The salient quality of contemporary society is its dynamic, fluid character. Today the conditions of life are profoundly different from those of a generation ago, and in all likelihood they bear very little resemblance to those that will govern your lives as professional men and women ten or twenty-five years from now. Day by day we make new discoveries and solve problems only to come upon another frontier, other possibilities, whole new areas of inquiry. We can merely speculate and surmise about what lies just over the horizon. But we can be quite certain that change will continue and that it will accelerate.

And so we shall hope to give you much more than the solid command of principles and facts without which there can be no mastery of a subject. Equally important, and even more lasting, will be the attitudes that you acquire, the perspective that you gain—a sense of history, an understanding of the nature of our society and of the forces that are shaping it. You will need a receptive mind, a willingness to expand your professional interests, the capacity to move with ease outside the area of your particular specialization, the confidence to draw upon and use with wisdom whatever may be new. To live and work fruitfully in this rapidly evolving, increasingly complex world you must develop the power and the ability to keep pace, to renew your knowledge. The true aim of education is to provide you with the inner intellectual and spiritual fiber to adapt—in sum, to teach you how to learn and how to think. Such, briefly, are the objectives that have molded our philosophy.

As you begin these four productive years, I can predict that your life—even as ours—will be intense and at times difficult. We shall make seemingly infinite demands upon your capacity to learn, and we shall set seemingly unattainable standards for your achievement. This is no place for idle minds or hands. You will probably work harder than you have ever worked before; yet never will you make a better investment of time and thought and energy.

Of course it will not all be study. You will participate in sports, in the varied activities, in student government. As individuals, and as members of our fraternities and dormitories, each of you in your own way can contribute constructively to the community to which you now belong. For all this, too, is an important part of education.

One day, some of you will be graduate students here, some will be members of our faculty, a few may be members of our Corporation. But all of you will be alumni and all of you will continue to be very much a part of M.I.T. Be proud of it, identify yourself with its welfare and progress as you do with your own. Taken as a whole, perhaps there is nothing quite like it anywhere in the world. For all of those who make it so, and for myself, I welcome you most warmly.

18

Charge to the Graduates

The closing words spoken at the Commencement Ceremonies at M.I.T. on June 8, 1962.

Our aim through the years that you have spent at M.I.T. has been to prepare you for your obligations and responsibilities as professional men and women in an age that is strongly dominated by the influence of science and engineering.

We have in fact set great store upon the achievement of professional excellence for its own sake. We have wanted to impart to you that intellectual self-reliance that rests upon a complete command of basic principles. We have constantly endeavored to impress upon you the need to view engineering and science as agents for the improvement of human welfare. We have tried to bring to you some understanding of the great world of ideas and philosophies that have moved men through the centuries. And finally we have hoped to convey to you our strong belief that knowledge and power must be guided by integrity of purpose and supported by the highest moral character.

You who graduate this morning draw upon the most extraordinary diversity of origins, of interests, and of talents; and there lies before you a comparable diversity of opportunity and experience. A great number of you will go on with advanced studies. Others will take up the active practice of your professions. Many of you will return now to your homes in foreign lands to assume positions of leadership and trust.

All of you have developed strong roots in science. Some of you will indeed become scientists; others, engineers. But from the graduates of this class will come also architects, lawyers, doctors, professors, economists, managers of industry—I shall predict also journalists, historians, men of letters. For the central thought that has shaped all our academic policies is that the education of an undergraduate at M.I.T. should be relevant to the needs and conditions of this age—should offer a foundation to any of the great professions of the modern world.

In these diverse careers we hope that you do supremely well. Every profession stands in need of the thorough technical competence that you have to give. You will bring to them your knowledge of the facts and the basic principles of contemporary science and engineering; you may contribute much from the methods that you have acquired, from the quantitative, analytic approach to problems that has become ingrained in your mode of thinking. But the professional estate implies much more than technical competence alone, and we shall have fallen far short of our aims if we have given to you nothing else. For the very fact that engineering and science have become so completely interwoven with the whole fabric of life in the twentieth century, you must be sensitive to the impact of your works upon the society of which you are members. You must bear your share of the responsibility for solving the human as well as the technical problems of our age.

The essence of that responsibility is a deep and abiding concern for the obligations of a citizen. Your country—whether it be the United States or a land abroad—needs not only the kind of advice but the kind of leadership that you have to give. The effort that you have given to your years at M.I.T., the hopes and often the sacrifices of your families, the knowledge and inspiration dedicated to you by this faculty, have been an investment in your country's future as well as your own. In return, we ask that you give willingly of your time, your creative thought, and your personal faith to solving the problems of a working democracy.

And so you begin now to put your education to the practical test. The courses, the grades, the thesis, have value and meaning only insofar as they have prepared you for what lies ahead. But I must warn you that you have by no means escaped from the stresses of quizzes and examinations; for life, if it is to hold out any challenge, is a perpetual sequence of examinations; and each of you throughout your life will continue to compile your own cumulative record. I most sincerely believe that what you have learned here, the product of these rigorous years, the ability to marshal your facts under pressure, to meet deadlines, to follow through to the end, your whole attitude and approach to new situations and new problems, will stand you in good stead. Only now will you be able to assess this education for its real worth; for the first time you will see M.I.T. in a new perspective. And M.I.T., in turn, will be judged by what you are and what you accomplish.

19

Three Men of M.I.T.

KARL TAYLOR COMPTON:
EDUCATOR AND ADMINISTRATOR*

On the afternoon of the twelfth day of March, 1930, in a brief and simple ceremony, Karl Taylor Compton was presented to our faculty convened in Huntington Hall as the ninth President of M.I.T.

Karl Compton came to us in a time of adversity. We had lost President Maclaurin at a most critical point in our history. There had been a long interregnum without a president. The postwar demands of industry upon our faculty, the chaotic effects of inflation in the twenties upon all academic life, and then the onset of the depression years—all these conspired against the progress of the Institute. With his coming, M.I.T. began to surge with new life and a new faith. No man ever fulfilled more completely his first promise; for much of the greatness of M.I.T., as you know it, is the fruit of his labors.

On that afternoon in the spring of 1930, Karl Compton was, to most of our faculty of engineers, little more than a distin-

* Words spoken at the Compton Memorial Convocation held at M.I.T. on October 4, 1954.

163

guished name. He was an outstanding physicist, already a noted figure in the scientific life of the country, yet to our faculty a stranger. In an incredibly short time all this was changed and we took him to us as one of our own. He became our friend, our confidant, and our leader. His roots were deep in academic soil. His father was in turn professor of philosophy, dean, and acting president of the College of Wooster. Both his brothers became presidents of colleges, and his sister was the wife of the president of a Christian college in India. No family has made more distinguished contributions to American education than the Comptons. Karl Compton too had served his apprenticeship: instructor in chemistry at Wooster, and then doctor of philosophy, *summa cum laude,* at Princeton; for three years instructor of physics at Reed College, followed by a return to Princeton as assistant professor of physics—eventually to become research professor and chairman of the Department. Teaching and research were in his blood.

It was no easy decision for Dr. Compton to abandon this life that he loved at Princeton and to assume the multitude of cares that burden every college president. We all take pride in the vast prestige achieved lately by the Institute, but one must not belittle the past in order to brighten the present. A quarter-century ago M.I.T. was an institution of world renown. We were no less thorough in our methods or less jealous of our standards than today; our objectives only were narrower, our field of interest more confined. In 1930 we were in essence an undergraduate school of engineering. Mathematics, physics, economics, and the humanities all were subservient to the requirements of the engineering curricula. Science, as Dr. Compton had known it, graduate study, and research were almost wholly absent. We belonged to the broad family of academic institutions but were of a species quite foreign to his earlier interests and experience.

You are all aware that M.I.T. was founded on a plan and philosophy of education formulated by William Rogers more than a century ago. Karl Compton understood that plan, adopted it, and was loyal to it throughout the remainder of his life. He sought not to change the aims and methods of the Institute but to enlarge and enrich them. All this he made very clear in his inaugural address with these words:

> There appears to be no reason for any change in the purposes and ideals of the Institute. It has been devoted in the most fundamental way to the benefit of mankind through science. There is every indication that only a beginning has thus far been made in the science of discovering and understanding nature, and in the art of usefully applying this knowledge. I can conceive of no more appropriate or urgent program than simply to continue the work of developing both principles and men for applying science to the problem of human welfare.

And then in that same first address he set forth the aims that would guide his own administration. I should like to read them to you in part. In June of 1930 they expressed his goal; twenty years later they might summarize his achievement.

> Although the purpose of the Institute is unaltered, I do believe that present conditions indicate the necessity of careful attention to several vital matters. First, I would suggest the necessity of greater emphasis upon the fundamentals of science, both in their own rights and as a basis of the various branches of engineering. As engineering has developed greater and greater complexities, it becomes increasingly impossible to hope to train men in those exact processes of thought or manipulation for which they will later be called upon. Many who start in as engineers later become executives or administrators. In all such situations a broad and full training in fundamental principles gives much greater power than a training in details which may seldom be encountered in practice. Whereas a generation ago most of our great technical industries were in their infancy and needed many men trained in the details of their respective arts, now most of these industries are large organizations which are equipped and prefer to train their own men in the fine points of their art: they absolutely require, however,

men who come with a sound basis of training in fundamental principles. The Institution which supplies these men, supplies the men destined to leadership.

I hope, therefore, that increasing attention in the Institute may be given to the fundamental sciences; that they may achieve as never before the spirit and results of research; that all courses of instruction may be examined carefully to see where training in details has been unduly emphasized at the expense of the more powerful training in all-embracing fundamental principles. Without any change of purpose or any radical change in operation I feel that significant progress can thus be made.

Second, let me emphasize the supreme necessity of maintaining a Faculty of absolutely first grade. The needs of an educational institution for the best men should supersede the claims of any other organization, for it is these men in the educational institutions who train and inspire all the others; their abilities are renewed and made available to the world in every graduating class.

These then were his views as he took over the reins of administration.

There are many men with ideas, but ideas without action serve little purpose. William Barton Rogers was a great innovator in education not simply because of the novelty of his ideas but also because he was able to express them in the establishment of a new institution. Karl Compton was one of the great leaders of our time not only because of the depth of his understanding of the needs of education in science and technology but because he met and fulfilled those needs through the institution entrusted to his care.

If you who are students of M.I.T. today find here first-rank departments of science, it is because Karl Compton knew the true qualities of science, believed that science must be cultivated for its own sake, and possessed the power to draw men of stature to his side.

If you observe here a school of engineering acknowledged throughout the world to be without peer, it is because Dr. Compton sought to nourish engineering education at its roots, to weed out elements that with time had become vocational or obsolete, and because he sought tirelessly for teachers of distinction.

If at M.I.T. you find a fusion of the humanities with science and technology in a manner that has begun to challenge the traditional patterns of liberal education, that is because he envisioned for the scientist and engineer a role in industry and government of steadily widening responsibility.

If, finally, you have sensed that with the faculty you are sharing in the life of one of the great research centers of our time, it is because of his conviction that teaching becomes sterile unless refreshed at the wellsprings of new knowledge.

Dr. Compton was a wise and skillful administrator, but it was by no simple art of administration that he brought M.I.T. into the small company of great universities. In every sense he was a leader, matching a fine mind with a radiant personality. As one looks back over the years of his presidency, it is easy to discern the tangible, material progress of the Institute; but the great gift that he bestowed upon us was his spirit, the shining example of his own life. We were moved by the transparent honesty of his aims. We felt the warmth of his interest in each and every one of us, and students and faculty alike, we responded with loyalty and pride in his own great achievements.

The fame that M.I.T. has achieved throughout the entire world has grown to such proportions that it frightens many of us who are so keenly aware of our own defects and of the long road that lies ahead before we can attain our ultimate goals. Yet this fame rests on the solid accomplishments of our graduates, upon the enormous contributions to the advancement of knowledge and the unselfish services made in the public interest by our faculty. And underlying these evidences of our new stature, there is a hidden strength that comes from an inner harmony, a unity of purpose and action, an extraordinary absence of factionalism and petty dissension.

This is why M.I.T. is to me, and I know to you, a good place to be. This was Karl Compton's spirit; this was his true bequest.

167

FRANCIS FRIEDMAN:
*TEACHER OF SCIENCE**

We who have come here this morning are the family, the friends, the colleagues of Francis Friedman. We loved and respected him, and we have wanted in this way to pay our personal tribute to his memory.

I recall very distinctly his first years at M.I.T. Following the war, great changes took place in our Department of Physics. Many new colleagues joined us at that time. Some were older men who already had made distinguished names: Victor Weisskopf, Bruno Rossi, Jerrold Zacharias, among others. But many were young men of outstanding promise who in one wartime activity or another throughout the country had attracted the attention of their seniors. Indeed, M.I.T. in those years set out to build upon youth.

Francis Friedman was one of these. He came to us in 1946 as Research Associate after four years in the Metallurgical Laboratory of the Manhattan Project at the University of Chicago. He began his career at M.I.T. under Jerrold Zacharias in the newly established Laboratory for Nuclear Science. He took an active part in a series of studies that made technical history in the United States during that period: the Lexington Project, Hartwell, particularly the Troy Project that brought him to the attention of a number of our senior faculty from fields other than physics. He waited until 1949 before completing his doctorate, and in 1950 he became a member of our faculty.

From the outset it was clear that this man was different. He seemed free from, or at least he resisted, the forces that drive so many of our brilliant, ambitious young scientists in the academic world today: the drive to publish, to find the problem with the quick pay-off, to distinguish oneself from the crowd. Francis

---

* Words spoken at a memorial service held at M.I.T. on August 8, 1962.

never did publish widely. And yet he came to be known and respected by leading physicists everywhere in this country and abroad. For he possessed extraordinary qualities of mind and character, a penetrating power of analysis, a most remarkable lucidity of thought, and above all he had taste, the ability to discriminate between the important and the trivial. He was interested in the whole wide range of science and the relations of science to all the other intellectual interests of mankind. No one who knew or worked with him could fail to sense this penetrating depth and judgment, even more impressive than the brilliant mind.

We waited—somewhat impatiently at times—for him to focus on his great work. He found it finally in education. I do not know when it began; probably it was always with him. But he was moved—almost to the point of obsession—by an enormous belief in the power of education; and for this he was willing and glad to dedicate his career and his life.

The Physical Science Study Committee—that extraordinary pioneering enterprise—gave him his first real opportunity. Everyone who participated will testify to his contribution. He was the rock upon which the project grew. He brought to it the intellectual integrity, the utter objectivity that marked everything he did. The work was illuminated by his own clarity of mind. Above all he gave to those studies his extraordinary insight into what was essential, the balance of emphasis between the important and the trivial. Out of that brilliance came a pedagogical insight, the intuitive recognition of what the student would easily understand and what would remain difficult for him to grasp.

From the PSSC his interests in problems of education began to widen. He became fascinated by the prospects and the opportunities for the teaching of science in the elementary schools—torn between this and his interest in college physics, which was one of the last things he and I talked about.

With Jerrold Zacharias he became much interested in problems of teaching in underdeveloped countries, particularly in Africa.

But his most pressing responsibility this past year has been in the development of our Science Teaching Center.

It has been our desire to capitalize on the experience of the PSSC, to turn these lessons to the advantage of our own students, and students in colleges everywhere: not to dictate curricula and courses but to provide the materials for the teaching of science and the insight and experience on effective methods.

We are very proud that this work had its inception here, that M.I.T. has from the beginning given support and encouragement. Indeed the life of Francis Friedman has been in unity with many of the hopes and aspirations of this institution: that we may contribute powerfully to the advancement of science but also that we shall be known far and wide for the leadership that we have given to education itself; that we are breaking new paths, demonstrating by our own example how best we may convey an understanding of science to the generations of students who come after us.

For all these hopes and plans the loss of Francis Friedman seems almost irreparable. But this is no time to think only of our loss. He had six wonderful, constructive years. He has left with us a treasure of ideas and objectives. His last wish was that this work go forward. There can be no better way to honor his memory than to carry on over the course he has set—and this indeed we shall do.

CHARLES STARK DRAPER:
AN APPRECIATION*

Just thirty years ago the teaching of physics at M.I.T. underwent a thorough revision. The changes were so drastic that they amounted in effect to the creation of an entirely new department. At the request of President Compton, a very youthful professor,

* The foreword to *Air, Space, and Instruments,* edited by Sidney Lees (New York: McGraw-Hill Book Company, 1963), a Draper anniversary volume.

John C. Slater, assumed the leadership. New members of a physics faculty arrived from Harvard, from Stanford, from Princeton, and from a number of other institutions. Within a remarkably short time, the undergraduate curriculum was completely overhauled, a graduate program of exceedingly high standards was established, and major research in modern physics was under way. Those of us who had the good fortune to participate recall a tremendously stimulating intellectual experience and a critical moment marking the serious entry of the Institute into the field of basic science.

My own particular assignment in this enterprise was to develop a series of graduate courses in analytical mechanics and fluid dynamics. The approach was wholly theoretical and classical, patterned upon the traditional models current in the leading graduate schools of the day, and strongly reminiscent of the Cambridge tripos of the preceding fifty years. Naturally, the principal texts were Whittaker, Lamb, and Love. But the work of Born and Heisenberg was already influencing classical theory as well as quantum mechanics, and from the first we put heavy emphasis upon the variational principles, the Hamilton-Jacobi equation, and the perturbation of orbits—all this without the stimulus of artificial satellites. This was also a period of rapid progress in aerodynamics. We soon found more life in Prandtl than in Lamb and ventured shortly into the more difficult and uncertain domains of viscous flow and boundary layer theory.

In the fall of 1931 I offered my course in Advanced Mechanics for the first time, followed by Hydro- and Aerodynamics in the spring. Stark Draper was a member of that class. It was the beginning of an association as friends and colleagues that has continued over the three decades of our careers at M.I.T. Clearly I can claim Stark as one of my former students. However, to make much of that point would be a travesty of the facts; it would be difficult to say whether it was teacher or student in the course of that winter who learned most from the other.

No one who has had Draper in a class is ever likely to forget

the experience. The first indication of a radically new phenomenon among students appeared in the assigned homework. There
is a particular type of brilliant student, known to every teacher,
whose genius is focused on economy of effort. He will solve the
most difficult problem in a few lines; at least the answer is correct,
although how he arrived so simply at the conclusion is clothed
in some mystery. An entire assignment is dealt with in a page or
two. Not so Draper. Every exercise to illustrate theory became
the subject of a thorough and painstaking investigation. If a problem caught his interest—and most problems did—the solution
took on the proportions of a thesis. The limiting assumptions
were examined, the special cases carried through to his satisfaction, and every physical implication analyzed to exhaustion. A
harassed instructor is disinclined to view such monumental pro-
ductions as an unmitigated blessing, but one shortly discovered
that Draper's discussion of classical exercises invariably afforded
a deeper insight into the subject than the original texts.

The most difficult task in the teaching of advanced theoretical
physics is to hold fast to the physical reality that underlies the
mathematics. Both instructor and students are commonly carried
away by the beauty of the formal structure and often part company completely with the material problem that the theory was
designed to solve. To deal with a situation that any freshman
might handle simply and accurately by the relation $F = ma$, the
graduate student is prone to marshal an arsenal of differential
equations. Once in a doctoral examination, with the charitable
intent of putting a confused candidate at his ease, I asked him to
solve an elementary problem involving a pulley, a weight, and a
block sliding on an inclined plane. I recall my shocked amaze-
ment when he proceeded to set up the appropriate Hamiltonian
function.

Such a blunder in method would never occur to Draper. Perhaps the outstanding quality of his mind lies in his ability to cut
through to the physical core of the problem. To many who have
worked with him in later years, this extraordinary ability to strike

to the root of the matter is so unfailing that it appears to be wholly the product of intuitive genius. Without questioning the genius which he clearly possesses, I think that this interpretation is unfair. For it fails to credit the countless hours of thought that he has given over the years to his chosen field, to the infinite pains with which he has examined the physical meaning of each problem. A great deal of intuitive insight is the consequence of just that kind of hard work. He would never construct a Lagrangian function when Newton's equations were amply sufficient for his purpose. The seemingly simple vector equation $\overline{M} = \overline{\omega} \times \overline{H}$ has been the key to his work on gyros. But when he has pressed these simple methods to their limits, he moves on into the domain of more complicated mathematics with ease and confidence.

This insistence on relating every theoretical problem to a real case and on applying the analysis to the physical essentials is characteristic of all Draper's work, and it contributed to one of the most interesting courses in my own teaching experience. Rigid-body dynamics is, of course, a standard part of every treatment of analytical mechanics. It is difficult, and unless one takes care, it can also be dull. Whether or not that year marked the beginning of Draper's interest in rotating coordinate systems, I do not know, but I do recall lively hours of discussion on the properties of moving and inertial reference systems. Under his prodding, we departed from the texts of Whittaker and Klein and Sommerfeld and looked pretty thoroughly into how a real gyroscope actually behaves.

In all this, Draper already had one material advantage over the rest of us. He was an accomplished pilot and constantly inquired how our theory might be applied to the current problems of the airplane. One of these, I remember, had to do with the phenomenon of ground looping, which in the early 1930's was plaguing the airlines. The problem in fact was a rather simple one, but the discussion conveyed to the class by vivid examples the nature of certain types of dynamic instability.

However, it was Stark's spectacular demonstration to me of

another kind of aerodynamic instability and of the shortcomings of airplane instrumentation that made the most indelible impression. With the rapid increase in blind flying, the need for safeguards against stalling followed by spinning had become extremely serious. Draper explained to me in detail the nature of the difficulty, how it could be avoided by proper procedure and instrumentation, and volunteered to give me a practical demonstration. I demurred. At length one Sunday I ran short of excuses and, lacking the simple courage to say no, took off with him in a small plane with an open cockpit. It was a beautiful morning; my confidence ebbed slowly back as we flew serenely over Concord and then headed toward Boston. I sat behind Stark as he lectured with shouts and gestures on the idiosyncrasies of instruments. Then, at a point directly over the harbor, it happened. The nose went sharply up. For a moment we hung on the propeller, as if suspended in the stillness of space. Suddenly the plane pitched forward; I gazed down on the water far below. We began to pick up speed. I became aware of a strange phenomenon: slowly, then more rapidly, the Boston skyline began to rotate about the plane. The Customs House tower moved in a majestic circle. It occurred to me that I had left a good many things undone and that the Department would be hard put to find someone to teach mechanics. The plane leveled out, rolled over, and settled down on a quiet course. I said nothing. No student but Draper has ever done such a thing to me before or since.

But it all added up. The qualities of mind and interest that students and colleagues have come to know so well were plainly evident in his own graduate years: the bulldog tenacity with which he shakes the most stubborn problem until the solution comes tumbling out; a complete command of the necessary theory, acquired as much through hard work as through native mathematical genius; a phenomenal sense of the physical implications of each problem; the constant concern to test theory by a specific, practical example. These powers he knits together into an effective driving force through a singular concentration of purpose. Stark Draper marked out the central domain of his interests a long while ago. He has pursued them faithfully and effectively for some thirty years. His success has distinguished him as one of the foremost engineers of our time.

Early events of World War II shaped his efforts along the particular course that has won him such great renown. In 1941, the United States was shocked by the sinking of the *Repulse* and the *Prince of Wales,* with all the implications for the future of naval power. Out of a discussion with the late Nathaniel McL. Sage, whom he deeply admired and respected, Draper was led to the development of the Mark 14 gunsight. Two years later, at the battle of Santa Cruz, *Battleship X* (U.S.S. *South Dakota*) gave a spectacular demonstration of its effectiveness.

From the Mark 14 gunsight, in turn, there came a whole new concept of fire control. During the 1940's and early 1950's, a school of thought about control and guidance grew up around Draper and his students. For twenty years the Instrumentation Laboratory at M.I.T. has been identified with his ideas and theories. Indeed, the Instrumentation Laboratory has provided a medium for the expression of his remarkable creative powers and the means to translate concepts into physical reality.

The tangible products of the Laboratory have been systems and devices—hardware, in crude terms—of incalculable importance to the security of our country. Yet despite his legitimate pride in these achievements, I am absolutely certain that Draper through the years has found his supreme satisfaction in the association with students of all ages. To many it has long been a mystery how and why he should have resisted the powerful lures of industry. But those who know him clearly understand that his unfailing attachment to M.I.T. stems from a passionate devotion to teaching and to the environment of an educational institution.

To Draper there has never existed a dichotomy between teaching and research; the two are of one and the same fabric. Nor has he bothered often to distinguish between the formal and informal processes of education. With Stark, teaching is something that goes on through all the waking hours, consciously or instinctively, in his office, at lunch, in the laboratory, at the colloquium, or in the lecture hall; discoursing with technicians, with his colleagues, with undergraduates, with naval officers assigned to his laboratory, with all and sundry who gather about to debate and learn. If he harbors a single discontent with his life at M.I.T., it is in the suspicion that perhaps from time to time the administration—and even a few colleagues on the faculty—have failed to grasp adequately the function and the potentialities of the Instrumentation Laboratory, from its position on the very frontiers of science and technology, as a superb base for the teaching of modern engineering.

The Laboratory is indeed part and parcel of Stark Draper's approach to engineering education. Through it he is able to impart tangible reality to theory and to vest it with importance. His influence upon several generations of students in the fields of instrumentation, guidance, and control has been enormous. They will be found in the ranks of industry, in the several branches of the military service, and in the engineering schools of the country. They are the finest testimony of a marvelously productive career.

M.I.T. shares in the pride of Stark Draper's achievements. I speak for his teachers and students of former years and for the multitude of his friends and colleagues in this warm tribute on his sixtieth birthday and in the good wishes we extend for the many years to come.

20

A Commitment to the Earth Sciences

Words spoken at the dedication of the Cecil and Ida Green Building (Earth Sciences) at M.I.T. on October 2, 1964.

As a geologist, William Barton Rogers took an intense interest in the rock and soil formations of New England. In fact, he located our first building at the edge of Copley Square in Boston on the soft, wet mud freshly pumped out of the Back Bay. With this decision he clearly established a precedent for others to follow; for although our records are silent on the subject, only a consuming confidence in the future progress of soil mechanics could have induced President Maclaurin and the Corporation to choose in 1912 this present site—a marshland of postglacial silt, clamshells, and fishweirs—for the foundations of the new M.I.T. That this magnificent building behind me rises straight and tall and firm is in itself no small tribute to the advance of engineering.

There was another decision in the history of the earth sciences at M.I.T. that came later—a conscious decision, a more philosophical one but crucial for our presence here today.

About fifteen years ago, not long after the war, we began a quiet reassessment of the goals and activities of the Institute. A whole array of glamorous developments was enticing us into new areas of endeavor—the new electronics, information theory, high-energy physics, nuclear engineering—all subjects of immense theoretical and practical importance. Moreover, with time, still others would appear, and many would demand a place in our academic portfolio. But we were conscious, too, of our own limitations in resources and of the need to focus our efforts and to set our future course wisely. And so we had to answer the question of whether or not the time had come for some pruning and cutting away, whether perhaps some older, classic fields—and geology was one of them—might now have to be abandoned.

Happily, out of those deliberations there emerged a full understanding of the enormous and continuing need to explore further and to use effectively the basic resources that nature offers to man and the environment in which he lives: the solid earth and its minerals, our supplies of water, the oceans and the life they contain, the atmosphere and its movements, and now outer space and the planets themselves.

For science, knowledge is an end in itself. But as the population of the world grows, as our resources of nature begin to dwindle, the knowledge that comes to us from this vast domain of learning can have also a supremely practical import for the future of mankind.

It is with this conviction that M.I.T. has reaffirmed its commitment to the earth sciences. It is with this objective that we have undertaken to add further strength to our efforts. And I know that it is in this same spirit that Cecil and Ida Green together have made what they are happy to call their "investment"—a magnificent testimonial of their concern for human welfare, of their faith in the power of science and engineering to stimulate progress, and of their confidence that M.I.T. will give reality to their hopes and desires.

One day last spring, when the lights first began to appear in the new building—the shakedown cruise, so to speak—Dean Fassett received an anonymous letter from a student. It read as follows:

Dear Dean Fassett:

Saturday night, or maybe Sunday morning, after a party, a companion and I were walking along the Esplanade on the Boston side of the Basin. The air was phenomenally warm, about 55 degrees, and the Charles was like a mirror; in this misty state of affairs we viewed something uncommonly beautiful, which I feel bears bringing to your attention. The new Earth Sciences building was illuminated from bottom to top and the effect of the lighted structure, with its toes touching its reflection in the misty millpond, was nothing short of spectacular! As we watched, several cars slowed to a stop on Storrow Drive to look.

My heartfelt thanks and appreciation to the man, whoever he is, who had the building lighted at that hour. . . . We'll show Boston that M.I.T. can be beautiful.

Sincerely yours,
A Sophomore Fraternity Man

We shall indeed.

Nothing that I might say could add to the simple sincerity of that tribute to the beauty of this building.

21

Looking Back

Words spoken at the dedication of the Student Center at M.I.T. in Dr. Stratton's name on October 9, 1965.

I remember as though it were only yesterday that evening of late summer when for the first time I walked across the old Harvard Bridge and saw the new buildings of M.I.T. looming white and stark across the Charles. I had come a long way, and I remember the first welling up of pride—that first sense of belonging.

Yet on that August night so long ago, nothing could have been further from my thoughts than the idea that I should ever be anything here but a student. I had other plans and other hopes, and a lifetime in school was not among them. I would stay a little while then go out into one or another of the careers that at that moment seemed to me so immensely appealing.

It was not to turn out that way. I was, in fact, destined to become a captive. Again and again, as the years went by, I came to the point of breaking away, of shaking off the ties to this campus and going out to explore new lands. But each time some strange, compelling force—something that to this day I find hard to explain—drew me back and held me here. And over the years, that curious mixture of feelings that most of us have known as undergraduates—that mixture of pride and respect and even occasional anger—grew into a consuming devotion to M.I.T.

As I look back over these four decades, I do not see how anyone could have asked more of life or fortune than to have had some part in this endeavor. I have been able to watch at firsthand the complete transformation of this institution—the vast expansion of our campus, the fine new buildings that surround us here today on what was in my own student days an utter wasteland, the renown of M.I.T. that reaches now to every part of the world. And all this tangible progress merely testifies to something more —the extraordinary vitality that pervades every part of the Institute, the sense of dynamic, constructive purpose, of onward movement.

To the excitement of living in such an environment, I have added the privilege of my friendships with one generation after another of students, of following their success, and of pride in their achievements as alumni. For in the final analysis, the student is the essential reason for our being.

What you have done for me this afternoon is the high and culminating point of my own years at M.I.T. Nothing that has happened before and nothing that can ever happen to me in the future can carry quite the meaning of just this moment. It is an enormous honor that a building should bear one's name. And that the naming of the Student Center—for me—should come as a spontaneous wish of students enhances that honor more than I can possibly tell you. Indeed, I find no words to say how deeply I have been moved or to express my gratitude. Surely you must understand that I do thank you with all my heart.

Index

183